Probability and Statistics

for use with

Second Canadian Edition

CALCULUS FOR
THE **LIFE SCIENCES**

Modelling the Dynamics of Life

Adler
Lovrić

Written by Miroslav Lovrić

NELSON

NELSON

Probability and Statistics
by Miroslav Lovrić

for use with *Calculus for the Life Sciences*, **Second Canadian Edition**
by Frederick Adler and Miroslav Lovrić

Vice President, Editorial Higher Education:
Anne Williams

Publisher:
Paul Fam

Executive Editor:
Jackie Wood

Marketing Manager:
Leanne Newell

Developmental Editor:
Suzanne Simpson Millar

Technical Checker:
Caroline Purdy

Content Production Manager:
Claire Horsnell

Copy Editor:
Heather Sangster at Strong Finish

Design Director:
Ken Phipps

Managing Designer:
Franca Amore

Cover Design:
Martyn Schmoll

Cover Image:
iLexx/iStockphoto.come

ISBN-13: 978-0-17-657135-1
ISBN-10: 0-17-657135-3

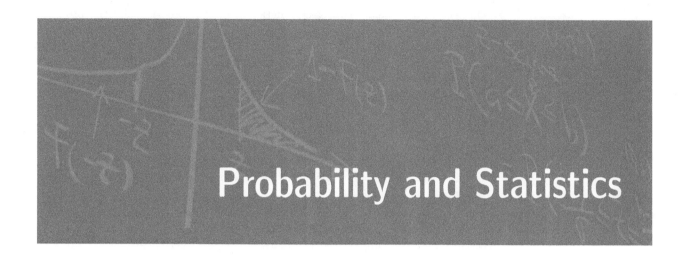

Probability and Statistics

The goals of this module are to provide a comprehensive and thorough introduction to the theory of probability, to present a variety of probabilistic models and applications, to expose the reader to basic forms of statistical thinking, and to build solid foundations for a serious study of statistics. Deeper understanding of statistics and the ability to apply statistical reasoning require a good understanding of the underlying probabilistic concepts and ideas.

In calculus and many other disciplines (within mathematics and beyond), we study idealized models. No matter how many times we solve a given differential equation, or run the exponential decay model, the answers are always the same. Departing from the deterministic view, we now focus on situations that involve chance factors, thus bringing us steps closer to a realistic context. Our analysis and conclusions about the processes that we study are based on data. But the real data are full of *noise,* which tends to obscure important features of the underlying process. Extracting *true,* valuable data from the real, noisy data is the realm of statistics.

Written with life sciences students in mind, this module develops all relevant concepts from the theory of probability and discusses a wide array of applications. A variety of approaches (algebraic, numerical, geometric, and verbal) facilitate the learning process and are aimed at improving students' skills in probabilistic thinking, reasoning about chance events, and communicating probabilistic and statistical ideas and information. Of course, we do not shy away from mathematics: all definitions and theorems are here, accompanied by fully solved examples, diagrams, and illustrations that help us understand things better.

What's in the module?

In Section 1 we make a formal transition from deterministic models to probabilistic (stochastic) models by introducing the concepts of chance and randomness. In parallel to the deterministic model, we outline the constituent parts of the stochastic model. The stochastic models that we discuss in Section 2 include animal population changes due to chance immigration, dynamics of disappearance and recurrence of a virus, analysis of chance in genetics, and a probabilistic view of the diffusion process.

The basic concepts that we work with—sample spaces, events, probability, conditional probability, the law of total probability, Bayes' theorem, and independence—are discussed in Sections 3, 4, and 5. As needed, we review definitions and formulas from set theory.

Random variables are introduced in Section 6. Deciding to study discrete random variables first, we postpone continuous random variables until Section 13.

In Section 6 we relate two basic functions to a random variable: the probability mass function and the cumulative distribution function. We measure the centre of a distribution in Section 7 and study its spread in Section 8. After introducing joint distributions (Section 9), we spend a good amount of time investigating important discrete distributions: the binomial distribution (Section 10), the multinomial and the geometric distributions (Section 11), and the Poisson distribution (Section 12).

In Section 13 we develop important concepts for continuous random variables. We briefly remind the reader of the basic integration techniques that are needed to work with probability density functions and cumulative distribution functions. Next, we study the normal distribution in detail (Section 14) and finish by presenting the continuous analogue of the geometric distribution, the exponential distribution (Section 15).

The approach used in writing this module—clear explanations and easy-to-understand narratives; numerous graphs, simulations, pictures and diagrams; a large number of fully solved examples and end-of-section exercises; and a wide spectrum of life sciences applications—makes the material suitable for students whose interests lie in life sciences and who are willing to deepen their understanding of life sciences phenomena.

I thank you for choosing this module, and I hope that you will like reading it and that you will learn some good and useful math.

Miroslav Lovrić
McMaster University, 2014

[Solutions to odd-numbered exercises from this module are posted (free download) on the web page www.nelson.com/site/calculusforlifesciences.]

1   Introduction: Why Probability and Statistics

By discussing a population whose dynamics is determined by **chance events,** we get our first glimpse into what **probability theory** is about. Collecting all outcomes from a **random experiment,** we apply **statistics** to analyze the outcomes and to communicate the results.

In this section, we talk about ideas and concepts mostly on an intuitive level. In the next section, we start building a rigorous theory that will help us understand chance and work with mathematical objects that involve chance.

Basic Definitions and Notation

The outcome of a *deterministic* experiment is certain. No matter how many times we repeat the experiment (for instance, solving a differential equation with an initial condition), the outcome is always the same.

We use discrete-time dynamical systems and ordinary and partial differential equations to describe deterministic events. We now revisit one dynamical system, and then we modify it in a novel way.

Example 1.1 **Basic Discrete-Time Dynamical System for a Population of Bacteria**

The number of bacteria in a colony with *constant* per capita production rate r changes according to

$$p_{t+1} = rp_t$$

where p_t is the number of bacteria at time t and p_{t+1} is the number of bacteria 1 time unit later (assume that the time unit is 1 hour). The number p_0 represents the initial number of bacteria in the colony. The per capita production rate is the number of offspring per individual bacterium.

Assume that $r = 1.05$ and that the initial population size is $p_0 = 200$. The deterministic system

$$p_{t+1} = 1.05p_t, \quad p_0 = 200$$

has solution

$$p_t = 200 \cdot 1.05^t$$

Thus, after 12 hours, there will be $p_{12} = 200 \cdot 1.05^{12} \approx 359$ bacteria; after 100 hours, $p_{100} = 200 \cdot 1.05^{100} \approx 26,300$. No matter how many times we repeat this calculation, these answers will always be the same.   ▲

In reality, it is a lot more likely that the per capita production rate will not remain constant but will fluctuate instead, due to various effects. Some we are aware of, such as the availability of food or the changes in ambient temperature. However, the population might be affected by events beyond our control, or by events that we are not even aware of.

We use the term *stochastic* to describe these unpredictable, unknown events (or effects). Often, we refer to them as *random* or *chance* events.

A *stochastic model* in life sciences describes a process that involves chance events. As in the deterministic case, the purpose of a stochastic model is to try to accurately describe the behaviour of the biological system that we are investigating.

Example 1.2   Stochastic Model for Population Growth

Assume that, as in Example 1.1, we start with 200 bacteria. However, this time the per capita rate is not constant—the population changes according to

$$p_{t+1} = r_t p_t$$

where the per capita rate assumes random values within the interval $[0.95, 1.15]$. This means that we do not know what the *exact* value of r_t is; all we know is that any number within the interval $[0.95, 1.15]$ is equally likely to be the per capita rate r_t.

This setup does sound more realistic. The bacterial population might have a "bad" time when it actually decreases (this is modelled by the fact that r could be less than 1) and a "good" time when it might grow by as much as 15% in an hour.

What will the population be 12 hours after the start of this experiment?

In Table 1.1 we record the per capita production rates and the corresponding population sizes for the first 12 hours. (For the per capita rates we used a random number generator programmed so that it picks any number in $[0.95, 1.15]$ with equal chance.)

Table 1.1

Time	Per capita rate	Population
0		200
1	0.9747	194.94
2	0.9868	192.37
3	0.9980	191.99
4	1.0335	198.42
5	0.9599	190.46
6	1.1305	215.32
7	1.1390	245.25
8	1.0482	257.07
9	1.0479	269.38
10	1.0175	274.09
11	1.1300	309.72
12	1.0328	319.88

In Figure 1.1 we compare our stochastic model with the deterministic model from Example 1.1.

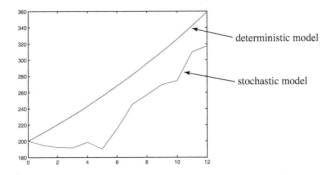

FIGURE 1.1

Stochastic and deterministic models for bacterial growth

What happens of we repeat the experiment? In Table 1.2 and in Figure 1.2 we show the outcomes of two more simulations.

Table 1.2

Time	Per capita rate	Population	Per capita rate	Population
0		200		200
1	1.1286	225.72	1.0985	219.70
2	1.0611	239.51	1.0349	227.37
3	0.9924	237.69	1.0359	235.53
4	1.1328	269.26	0.9750	229.64
5	1.0616	285.85	0.9549	219.28
6	0.9832	281.05	1.0080	221.03
7	1.1476	322.53	1.0135	224.01
8	1.0016	323.05	1.0807	242.09
9	0.9648	311.68	1.1414	276.32
10	1.0305	321.19	1.1371	314.20
11	1.0304	330.95	1.0416	327.27
12	0.9809	324.63	0.9981	326.65

We have just come across an example of a *random experiment*, i.e., an experiment that is *repeatable* but whose outcome is *uncertain*.

How do we report an outcome of a random experiment? So far, we have run it three times and the answers (for the population after 12 hours) are 320, 325, and 327. Suppose that we run the experiment 500 times. Listing the values of all 500 outcomes would not make much sense.

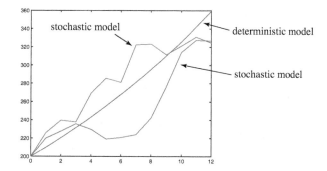

FIGURE 1.2

Two more simulations of the stochastic population model

Although the outcome of a random experiment is uncertain, this does not mean that we cannot quantify it. In Figure 1.3 we compare two more stochastic growth simulations (this time, run for 60 steps) with the deterministic solution. The simulations have been generated using random numbers, but their shape still resembles an exponential function. So how uncertain are the outcomes?

Clearly, what we need is a way to accurately describe all possible outcomes of a random experiment; as well, we need to measure and quantify uncertainty. This is what statistics and probability are all about.

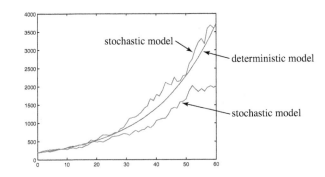

FIGURE 1.3

Stochastic versus deterministic growth

Definition 1   Statistic

A *statistic* is a set of numerical values that summarize the outcomes of a random experiment.

A statistic can be represented numerically, in the form of a table, as a graph, or as a diagram. What statistic we calculate, and how we represent it, depends on the questions that we need to answer.

For instance, one statistic we could relate to the outcomes of the 500 repetitions of a stochastic growth model is the mean (also referred to as the average value). Or, we can list the highest and the lowest values. In some cases, we might need to identify the outcome that is larger than 75% of all outcomes.

We go back to the stochastic growth model of Example 1.2. Although we do not know what a particular outcome will be, we know an interval within which all outcomes have to fall. Since all per capita rates are in $[0.95, 1.15]$, the lowest outcome occurs when the per capita production rate is equal to 0.95 during every time interval. In that case, the population count is $200 \cdot 0.95^{12} \approx 108.07$. The highest outcome, $200 \cdot 1.15^{12} \approx 1070.05$, occurs when all per capita production rates are equal to 1.15. Thus, we know that the outcomes of random simulations (Figure 1.4) have to fall between 108.07 and 1070.05. (Of course, we round off to the nearest integer when reporting on the actual count of bacteria.)

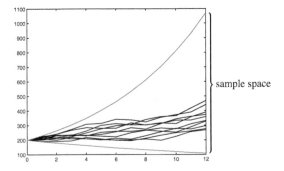

FIGURE 1.4

The range of all outcomes of the stochastic growth model

Definition 2   Random Experiment and Sample Space

A *random experiment* is an experiment that is repeatable but has an uncertain outcome. The set of all possible outcomes of a random experiment is called the *sample space* of that experiment.

In the stochastic growth experiment, the sample space consists of all real numbers between 108.07 and 1070.05. Using S to denote the sample space, we write $S = [108.07, 1070.05]$.

Are all values in S equally likely to occur as outcomes of the experiment?

To try to answer this question, we run the experiment 500 times, break down the sample space into intervals, and record the number of outcomes that fall within each interval. In this way, we construct the statistic called the *histogram;* see Figure 1.5.

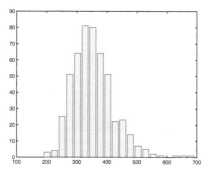

FIGURE 1.5

A histogram of 500 outcomes of the stochastic growth model

The sample space is placed on the horizontal axis, broken down into subintervals whose ends are at 201.2, 226.4, 251.7, 276.9, 302.1, 327.3, 352.6, 377.8, 403.0, 428.2, 453.4, 478.7, 503.9, 529.1, 554.3, 579.5, 604.7, 630.3, 655.2, and 680.4 (there were no outcomes smaller than 201.2 or larger than 680.4). The heights tell us how many outcomes fall within each interval. For the record, the heights are 3, 4, 25, 51, 64, 81, 80, 64, 51, 22, 23, 14, 7, 5, 2, 1, 0, 1, 1, and 1.

Note that although the per capita rates have been chosen with equal chance within the interval $[0.95, 1.15]$, the outcomes do not share the same property — some are more likely to occur than others. One of the major roles of stochastic modelling is to correlate the two: how does the randomness in the input affect the randomness in the output?

Example 1.3   Questions

We built a stochastic model that predicts (in its own way) the bacterial population size after 12 hours. We identified the sample space and drew a histogram that helps us answer some questions (such as finding which outcomes are more likely to occur).

This is just the beginning. There are many questions we might need to know answers to. How likely is it that the population will reach 90% of its theoretical maximum of 1070 bacteria? How likely is it that the population will fall between 200 and 350? We run the model 500 times and calculate the mean (average) value. How likely is it that 320 of 500 outcomes will fall within 15% of the mean?

As we learn probability and statistics, we will be able to answer these and many other questions.

Example 1.4   Stochastic Model for a Population with Large Fluctuations

Of course, we can run the model from Example 1.2 with different values.

To model larger fluctuations, we use the interval $[0.75, 1.35]$ for possible per capita production rates, instead of the interval $[0.95, 1.15]$. So, consider

$$p_{t+1} = r_t p_t, \quad p_0 = 200$$

where r_t is a number chosen randomly from the interval $[0.75, 1.35]$.

In Figure 1.6 we show the outcomes of several experiments. The magnitudes of the changes from hour to hour are a lot larger than before. This time, the sample space is the interval $[200 \cdot 0.75^{12}, 200 \cdot 1.35^{12}] \approx [6.3, 7328.8]$.

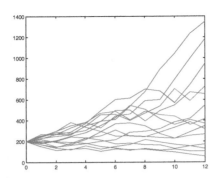

FIGURE 1.6

Simulation of a population
with large fluctuations

Definition 3 Stochastic Model

A *stochastic model* is a mathematical model that describes processes (such as
biological processes) that are driven by chance (random) events.

Stochastic models may contain both deterministic and stochastic factors. In Figure
1.7 we show what is involved in building such models.

We use *probability theory* to define the model and to obtain outcomes (results). Next, we use *statistical tools* to analyze and organize the outcomes and
then *descriptive statistics* to communicate them.

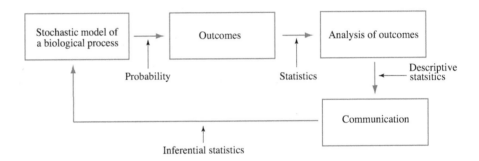

FIGURE 1.7

Stochastic model

In many situations, we do not stop once we have communicated the results
and our analysis of the results—we try to make inferences (deductions) about the
biological processes that we have been investigating. *Inferential statistics* provides
us with the tools necessary to make these deductions. For instance, determining
whether one population of bacteria grows more rapidly than another (say, based
on situations similar to what we did in Examples 1.2 and 1.4) requires inferential
statistics.

Probability theory was invented to analyze games of chance (gambling), but it
was soon recognized for its true strength—the ability to provide a precise mathematical description of chance (randomness). Using the theory, we are able to
answer all kinds of questions that involve chance.

Example 1.5 More Questions

Assume that there is a 60% chance that it will rain on Saturday and a 50% chance
that it will rain on Sunday. What is the chance that it will rain on the weekend?
Clearly, it's not going to be $60\% + 50\% = 110\%$. Do we have enough information
to answer this question? If not, what else do we need to know?

It has been determined that 0.9% of women over the age of 20 have breast
cancer. A commonly used test for the detection of breast cancer has a false-positive
rate of 10% (i.e., the test result is positive although a woman does not have breast
cancer) and false-negative rate of 3% (the test result is negative when a woman

actually has breast cancer). A woman undergoes the test and its result is positive. What is the chance that she has breast cancer?

Blood tests show that both a mother and a father are carriers of some trait (for instance, a genetic disorder such as high blood pressure). It is known that there is a 2% chance that the mother will pass the trait to the child and a 5% chance that the father will pass on the trait. What is the probability that their child will be born with that genetic trait?

We need to learn probability and statistics in order to understand uncertainty, chance, and risk, as they are an integral part of our daily lives. For many, it is not just understanding — health science professionals (among many other) need to *effectively communicate* the information about chance and risk to their patients. The extensive research (not just in health sciences) is the best evidence that the communication of quantitative information is far from trivial. To illustrate some issues involved, we look at a couple of examples.

The information about the side effects of Prozac (widely used medication for treatment of major depression, bulimia, panic disorder, and other conditions) states that there is "between 30% and 50% chance of developing a sexual problem." A large number of patients interpret this sentence incorrectly as "something will go wrong in 30% to 50% of my sexual encounters." [Source: Gigerenzer, G., Reckoning with Risk: Learning to Live with Uncertainty. Penguin Books, 2002]. Surveys show that this information can be understood better if we use relative frequencies, and say that "in a group of 10 people, between 3 and 5 will develop a sexual problem."

The statement "Event A happens in 5% of cases" is equivalent to the statement "Event A does not happen in 95% of cases." In other words, the two statements convey (mathematically) the same information. However, this cold and objective mathematical truth ceases to be what it is in real-life situations, such as the following. Two patients need to decide whether to undergo a surgery for a serious medical condition. Research shows that the patient who is told that there is a 5% risk of complications leading to death is more reluctant to undergo surgery than a patient who is told that he has a 95% chance of surviving the surgery.

In the next section, we will look at a few more models that involve chance and then start developing the theory of probability.

Summary   A **stochastic model** describes a process that involves unpredictable or often unknown **random** events (also called **chance** events). Using randomness, we were able to describe the fluctuations in the size of a population. A **random experiment** is an experiment that we can repeat as many times as desired, but whose outcome we cannot predict. A **sample space** is the set that contains all possible outcomes of a random experiment. A **statistic** is a way to describe the outcomes of a random experiment. As we have seen, statistics can be presented as a set of numbers, in the form of a table, or as a **histogram**.

2	Stochastic Models

We examine several stochastic models to further illustrate the concepts introduced in the last section: the dynamics of a population with immigration, the disappearance and occurrence of a virus, randomness in genetics, and diffusion. We will use these models (and others) in the forthcoming sections to illustrate new definitions, ideas, and formulas as we introduce them.

Examples of Stochastic Models

A stochastic model might contain both deterministic and stochastic terms, as the following example shows.

Example 2.1   Lion Population with Immigration

Consider the following stochastic dynamical system for the population p_t of lions living in a certain region:

$$p_{t+1} = 0.95p_t + I_t$$

where

$$I_t = \begin{cases} 12 & \text{with a 50\% chance} \\ 0 & \text{with a 50\% chance} \end{cases}$$

and $p_0 = 160$. The time unit is 1 year.

The deterministic part $p_{t+1} = 0.95p_t$ is the usual exponential law with a per capita production rate of 0.95. The term I_t represents possible yearly immigration, i.e., an influx of new lions (say, due to the lack of prey in surrounding areas). The lions that migrate to the region join the lions living there and stay there.

It is not certain that new lions will join the exisiting population; it is only with a 50% chance that a group of 12 lions will migrate into the region within any given year. Without the immigration, the existing lion population will be decreasing, since its per capita production rate is below 1. Can a chance event (immigration) save it from going extinct?

Let's do some calculations. The initial population is 160. To simulate randomness, we flip a coin, adopting the rule that heads means immigration (influx of 12 new lions) and tails means no immigration. Suppose we flip a coin and the outcome is tails. In that case, $I_0 = 0$ and

$$p_1 = 0.95p_0 + I_0 = 0.95(160) + 0 = 152$$

We toss the coin again, and the outcome is heads. Thus, $I_1 = 12$, and

$$p_2 = 0.95p_1 + I_1 = 0.95(152) + 12 = 156.4$$

i.e., 156 lions. It's tails on the third toss, and so $I_2 = 0$ and

$$p_3 = 0.95p_2 + I_2 = 0.95(156.4) + 0 = 148.6$$

We continue in the same way—see Table 2.1, where we have recorded the population for the first 10 years.

In Figure 2.1 we show the outcome of this experiment, as well as two more runs of the same experiment. (This is an example of a random experiment—it is repeatable (we can do as many simulations as we like), but the outcome (the number of lions after 10 years) is unpredictable.)

What is our prediction for the population? Looking at the three outcomes in Figure 2.1, we cannot be sure about numbers, but it seems that the population will decline.

Table 2.1

Time	Immigration	Population
0		160
1	No	152
2	Yes	156.4
3	No	148.6
4	No	141.2
5	No	134.1
6	No	127.4
7	No	121.0
8	Yes	127.0
9	No	120.6
10	Yes	126.6

To get a more reliable prediction, using a computer, we run 1,000 simulations for 10 years, and obtain an average of 140.8 lions—confirming the hint we got from the three initial simulations. We conclude that random immigration will not be able to help the local population of lions keep or increase their numbers.

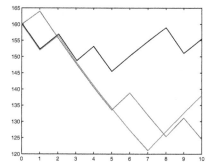

FIGURE 2.1

Three predictions for the lion population in 10 years

The histogram in Figure 2.2a shows what the most likely population counts will be, based on our simulation. Although the population will likely decrease, it will not go extinct, at least not any time soon.

To see what happens farther into the future, we run 100 simulations for 50 years, and obtain an average of 122.2 lions. The histogram of the 50-year predictions (Figure 2.2b) shows that the population predictions shift toward lower numbers, but not in a dramatic way.

FIGURE 2.2

Histograms predicting the population of lions after 10 years (a) and 50 years (b)

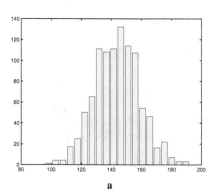

a

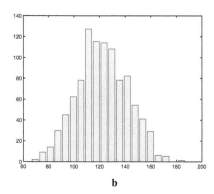

b

Example 2.2   **Disappearance and Recurrence of a Virus**

Certain health conditions (such as viral infections) appear and disappear over time with seemingly no pattern. After an absence of three months, an infection might reappear within a population, persist for several months, and then disappear again.

This (quite common) situation can be modelled in the following way. If, at time t, the virus is present in some population, we define $I_t = 1$; otherwise, define $I_t = 0$. The unit of time is 1 month. Suppose that the following is known about the virus. If it is present in the population at this moment, it will be present the following month with a chance of 75%. If it is absent from the population at this moment, the virus will be absent from the population the following month with a chance of 80%. We visualize the dynamics in the diagram in Figure 2.3.

FIGURE 2.3

Disappearance and recurrence of a virus

Using a random number generator, we can simulate the outcomes of this situation. Starting with $I_0 = 0$, we run the simulation for 50 months; the outcomes are shown in Figure 2.4. Note the interchanges between shorter and longer periods of absence and presence of a virus.

FIGURE 2.4

Dynamics of the virus for the first 50 months

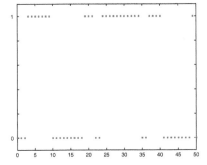

The system that we introduced in Example 2.2 is an example of a *Markov chain*.

Markov chains are characterized by the fact that the dynamics depends on the population. In other words, what the system (in our case the virus) does at a particular moment depends on the value of I_t at that moment. This was not the case with the lions in Example 2.1: the per capita production rate of 0.95 and the immigration of 0 or 12 are independent of the number of lions living in the region.

Example 2.3   **Chance in Genetics**

Many traits (in humans, animals, or plants) are determined by a pair of genes, one coming from the mother and the other coming from the father. In humans, traits include colour-blindness/no colour-blindness, lactose tolerance/intolerance, eye colour, right-/left-handedness, susceptibility to high blood pressure, and so on.

We assume that a gene comes in two variants (alleles); call them allele A and allele B. Each offspring inherits one allele from the mother and one allele from the father, and thus there are four possible combinations (called *genotypes*); see Table 2.2. The first letter is the allele from the mother and the second letter is the allele from the father. The two combinations AB and BA are viewed as the same genotype.

Table 2.2

Genotype	Name
AA	A-homozygotes
AB	heterozygotes
BA	heterozygotes
BB	B-homozygotes

Assume that both parents are heterozygous. Their offspring will have one of the four combinations AA (allele A from the mother and allele A from the father), AB (allele A from the mother and allele B from the father), BA (allele B from the mother and allele A from the father), and BB (allele B from the mother and allele B from the father). In the case where there are no outside factors influencing the selection of alleles, the four outcomes are equally likely. Thus, the offspring has a 50-50 chance of being heterozygous.

Assume that the mother's genotype is AA and the father's genotype is AB. Their offspring will have one of the following combinations: AA, AA, AB, and AB. Again, their offspring has a 50-50 chance of being heterozygous. ◤◣

Example 2.4   **Chance in Genetics II**

Assume that a plant is diploid (has two copies of each gene); we name the variants (alleles) A and B. Starting with a heterozygous plant, what is the chance that its offspring in generation t $(t = 1, 2, 3, \ldots)$ will be heterozygous?

Let c_t be the chance that an offspring in generation t is heterozygous. Initially, $c_0 = 1$ (i.e., 100%, since we start with a heterozygous plant). As in the previous example, the offspring could be homozygous (AA or BB) or heterozygous (AB or BA). Thus, two of the four combinations are heterozygous, and so the chance is $c_1 = 0.5$.

A homozygous plant will produce homozygous offspring. A heterozygous plant will produce a heterozygous offspring in 50% of the cases; thus, $c_2 = 0.5c_1$. In general,

$$c_{t+1} = 0.5c_t$$

Since the solution of this (deterministic!) system is $c_t = 0.5^t$, we conclude that the number of heterozygous plants will decrease exponentially. ◤◣

Example 2.5   **Chance in Genetics: Phenotypes**

In general, there are three genotypes: the combination AA exhibits one trait, the combination BB exhibits another trait, and the combinations AB and BA exhibit the third trait. If the heterozygotes exhibit the same trait as the homozygote AA, then A is called the *dominant allele* and B is called the *recessive allele*. The dominant allele and the recessive allele form the *phenotype* of an individual. Thus, there are three genotypes, but only two phenotypes (A dominant or B dominant). Consider an example.

Assume that an individual who has a copy of allele A has unattached earlobes (dominant trait) and the with two copies of allele B has attached earlobes (recessive trait). One thousand couples are sent to colonize a distant planet. Assuming that within each couple there is a person with genotype AA (unattached earlobes) and a person with genotype BB (attached earlobes) what is the chance that their grandchildren will have attached earlobes?

▶ Since an offspring of genotype AA and genotype BB parents can only be of genotype AB, all children in the first generation will be heterozygous (and so will have
unattached earlobes). An offspring of an AB and AB couple can have any of the
four combinations AA, AB, BA, and BB. In three of the four cases, a copy of the
A allele is present. Equivalently, in one of the four cases the offspring will have
a BB combination. Thus, there is a 25% chance that the grandchildren will have
attached earlobes.   ▲

Example 2.6   Diffusion as a Stochastic Process

For various reasons, we study diffusion at two different levels. Using the partial differential equation $c_t = \sigma c_{xx}$ for the concentration of the substance that is
diffusing, we look at it on a *macroscopic* level.

Now we look at the diffusion *microscopically*—what do individual molecules,
or a small number of them, do? This is a real situation—various toxins and
enzymes can affect the behaviour of a cell even if present in very small quantities
(a dozen molecules will suffice).

We will soon see that there is a big difference in the probability of events if
we consider small and large quantities. It is a lot more likely that to toss 5 heads
in a row than 100 heads in a row. A small number of molecules is more likely
to exhibit a certain behaviour. It could happen that a dozen molecules move to
the same region within a cell by random motion—this is something that's nearly
impossible for billions of water molecules within the cell to accomplish.

In our first model we consider a single molecule located inside a fixed region
(such as a cell). Assume that, in a given time interval (say, 1 hour), there is a
15% chance that the molecule will leave the region. Once it leaves the region, the
molecule does not come back.

In Table 2.3 we model the behaviour of three molecules (as usual, we need a
way to generate randomness; flipping a coin will do here).

Table 2.3

Time	Molecule 1	Molecule 2	Molecule 3
0	In	In	In
1	In	Out	In
2	In	Out	In
3	In	Out	In
4	In	Out	In
5	In	Out	Out
6	In	Out	Out
7	Out	Out	Out
8	Out	Out	Out

How do we calculate the chance that a molecule is still within the region after 8
hours?

As usual, we build a dynamical system. Denote by p_t the chance that the
molecule is still inside the region during the time interval t. Thus, $p_0 = 1$ (i.e.,
100%, since we assume that initially a molecule is inside the region). After one
hour, the molecule is still inside the region with a chance of 85%. Thus, $p_1 = 0.85$.

After two hours, the molecule is still inside the region if it was inside the region during the first and second hours. Thus,

$$p_2 = 0.85 \cdot 0.85 = 0.85p_1$$

(Soon, we will state precise reasons why we multiply the chances, as well as make exact assumptions that will tell us when we are allowed to do so. For now, we reason intuitively: assume that we toss a coin twice—the chance of tossing heads both times is 25%, which is the product of the chances of heads in a single toss.)

Continuing in the same way, we obtain the *deterministic* dynamical system

$$p_{t+1} = 0.85p_t$$

whose solution is given by

$$p_t = p_0 \cdot 0.85^t = 0.85^t$$

Thus, the chance that the molecule is still inside the region after 8 hours is $p_8 = 0.85^8 \approx 0.2725$, i.e., 27.25%.

In other words, if there were 1,000 molecules inside the region initially, we'd expect to see about 273 still inside after 8 hours.

Example 2.7   **Example 2.6 Continued: Simulated and Predicted Numbers of Molecules**

Assume that there are 100 molecules in the region. After 1 hour there will be $p_1 = 100 \cdot 0.85 = 85$ left. After another hour, there will be $p_2 = 85 \cdot 0.85 = 72.25$ (72 molecules left). Note that

$$p_2 = 85 \cdot 0.85 = 100 \cdot 0.85 \cdot 0.85 = 0.85 \cdot (100 \cdot 0.85) = 0.85p_1$$

(This is an alternative to our reasoning in the previous exercise; perhaps it shows in a more transparent way that we need to multiply to move to the next hour.) Continuing in the same, way, we obtain $p_8 = 100 \cdot 0.85^8 \approx 27.25$, i.e., 27 molecules. Figure 2.5a shows the exponentially decreasing graph of p_t compared with a simulation done on 100 molecules.

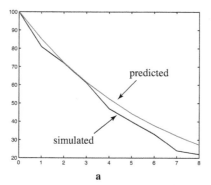 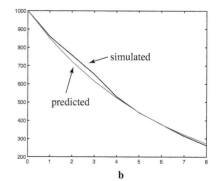

a   b

FIGURE 2.5

Simulated and predicted outcomes for 100 and 1,000 molecules

If we increase the number of molecules, the simulated and the predicted outcomes will get closer to each other. In Figure 2.5b we show the comparison for 1,000 molecules.

Example 2.8   **Modelling Diffusion Using Random Walk**

Diffusion is defined as the spread of particles from regions of higher concentration toward regions of lower concentration by means of random motion. Colliding with neighbouring particles, a particle constantly changes its direction, moving around in a random way.

We now describe a way of simulating this random motion. Assume that a particle is released from a source located at $x = 0$ on a number line and can move only along the line (we are looking at *one-dimensional diffusion*). During every time interval, a collision forces a particle to move either left or right with equal chance for 1 unit of distance.

A particle starts at $x = 0$ at $t = 0$. It collides with another particle, and at $t = 1$ with a 50% chance it moves to $x = 1$ and with a 50% chance it moves to $x = -1$; see Table 2.4.

Table 2.4

Location	Chance
−1	1/2
1	1/2

Assume that the particle ended at $x = -1$ at $t = 1$. The next collision moves the particle to either $x = -2$ or $x = 0$ with equal chance. If it ended at $x = 1$ at $t = 1$, the next collision will move the particle to either $x = 0$ or $x = 2$ with equal chance; see Figure 2.6.

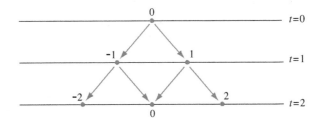

FIGURE 2.6

First two steps of the random walk along the line

Look at the situation at $t = 2$. There is one way the particle can end at $x = -2$ (moving left both times) and one way it can end at $x = 2$ (moving right both times). But there are two ways it can end at $x = 0$: moving left then right, or moving right then left. The chances that the particle ends at a particular location are given in Table 2.5.

Table 2.5

Location	Chance
−2	1/4
0	1/2
2	1/4

Consider another step. In Figure 2.7 we indicate the locations where the particle could end at $t = 3$. The numbers in squares represent the number of ways the particle can arrive at a particular location.

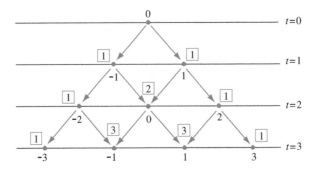

FIGURE 2.7

Random walk, third step

Of the total of eight paths, one leads to $x = -3$ and one to $x = 3$. Three paths each lead to $x = -1$ and $x = 1$. For instance, to reach $x = -1$, the particle can go left-left-right, or left-right-left, or right-left-left. The chances that the particle ends at a particular location are shown in Table 2.6.

Table 2.6

Location	Chance
−3	1/8
−1	3/8
1	3/8
3	1/8

We know that we cannot answer the question about the *exact* location of the particle after t collisions. But we can try to answer questions such as: where is the particle most likely to be? Or, how far from the source (from $x = 0$) is it most likely to be?

Take $t = 30$ (i.e., 30 collisions). We run the experiment for 30 random walk steps and then repeat 1,000 times. In Figure 2.8a we drew a histogram, showing how many times the particle landed at each location. Does the shape look familiar?

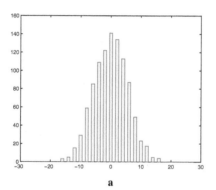 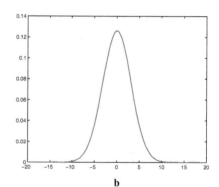

a   b

FIGURE 2.8

Random walk simulation compared to the Gaussian distribution

In calculus, we show that the concentration at a location x units from the source at time t is given by

$$c(x,t) = \frac{1}{4\pi Dt}e^{-x^2/(4Dt)}$$

where D is a constant (diffusion coefficient); see Figure 2.8b.

This particular shape is called the *Gaussian distribution function* or the *Gaussian density*. We will learn a lot about it later in this book.

Summary   Incorporating **chance factors,** we built models for a number of applications. All models involve **random experiments,** i.e., repeatable experiments whose outcomes we cannot exactly predict. We examined how a population of lions reacts to a possible yearly immigration and the dynamics of the disappearance and occurrence of a virus. Using chance, we explained basic principles in genetics. The concept of random motion helps us to accurately describe certain aspects of the process of diffusion.

2	Exercises

1. Consider the stochastic system for a population of lions given by $p_{t+1} = p_t + I_t$, where p_t represents the number of lions in year t, $t = 0, 1, 2, \ldots$. The immigration term is $I_t = 6$ with a 50% chance and $I_t = 0$ with a 50% chance. Assume that $p_0 = 100$.

    (a) Explain what dynamics is implied by this system. What would you expect the population to be 10 years later; i.e., what are the most likely values for p_{10}?

    (b) Using a coin to generate randomness, run the simulation three times, starting with $p_0 = 100$, and compare the values for p_{10} that you obtain.

    (c) What is the sample space for p_{10} (i.e., the set of all possible values for p_{10})?

2. Consider the stochastic dynamical system $m_{t+1} = r_t m_t$, where $m_0 = 1$ and $r_t = 2$ with a 50% chance and $r_t = 1/2$ with a 50% chance ($t = 0, 1, 2, \ldots$).

    (a) What is the chance that $m_2 = 1$?

    (b) What is the sample space for m_3 (i.e., the set of all possible values for m_3)?

    (c) What is the sample space for m_4?

3. Consider the stochastic dynamical system $m_{t+1} = r_t m_t$, where $m_0 = 1$ and $r_t = 2$ with a 50% chance and $r_t = -1$ with a 50% chance ($t = 0, 1, 2, \ldots$).

    (a) What is the chance that $m_2 = 1$?

    (b) What is the sample space for m_4 (i.e., the set of all possible values for m_4)?

4. Consider the stochastic system for a population of lions given by $p_{t+1} = p_t + I_t$, where p_t represents the number of lions in year t, $t = 0, 1, 2, \ldots$. The immigration term is $I_t = 3$ with a 50% chance and $I_t = -3$ with a 50% chance. Assume that $p_0 = 100$.

    (a) Describe the dynamics implied by this system.

    (b) What is your prediction for the population of lions in 10 years? In 20 years?

    (c) Using a coin to generate randomness, run the system three times and compare the values for p_{10} that you obtained.

    (d) How can you relate this system to the random walk discussed in Example 2.8?

5. A population of leopards p_t, $t = 0, 1, 2 \ldots$, is modelled by $p_{t+1} = p_t + I_t$. The immigration term is equal to $I_t = 3$ with a 75% chance and $I_t = -3$ with a 25% chance. Assume that $p_0 = 100$.

    (a) Describe the dynamics implied by this system.

    (b) What is your prediction for the population of leopards in 10 years? In the long term?

    (c) Using a deck of cards to generate randomness (how?), run the system three times and compare the values for p_{10} that you obtained.

6. A population of leopards p_t, $t = 0, 1, 2 \ldots$, is modelled by $p_{t+1} = p_t + I_t$. The immigration term is equal to $I_t = 10$ with a 90% chance and $I_t = -100$ with a 10% chance. What is more likely to happen to the number of leopards—an increase or a decrease? Or will the population remain at about the same size? Explain why.

7. Consider the following modification of the immigration pattern of the system $p_{t+1} = 0.95p_t + I_t$ in Example 2.1: $I_t = 12$ with a 75% chance and $I_t = 0$ with a 25% chance.

    (a) Explain how to simulate the probabilities that are needed in this exercise. [Hint: A deck of cards will do.]

    (b) Starting with $p_0 = 160$ lions, find the values of p_t for $t = 1, 2, \ldots, 6$. How does p_6 compare with the values of the three simulations shown in Figure 2.1?

8. Consider the following modification of the immigration pattern of the system $p_{t+1} = 0.95p_t + I_t$ in Example 2.1: $I_t = 12$ with a 75% chance and $I_t = -6$ with a 25% chance. Starting with $p_0 = 160$ lions, and using appropriate tools (see (a) in Exericse 7), find the values of p_t for $t = 1, 2, \ldots, 6$. How does p_6 compare with the values of the three simulations shown in Figure 2.1?

9. Consider a diploid plant of genotype AB.

   (a) What fraction of first-generation offspring has genotype BB?

   (b) What fraction of second-generation offspring has genotype BB?

   (c) What fraction of third-generation offspring has genotype BB?

10. Consider a diploid plant of genotype AB.

    (a) What fraction of first-generation offspring has genotype AB?

    (b) What fraction of second-generation offspring has genotype AB?

    (c) What fraction of third-generation offspring has genotype AB?

11. The trait of the dominant allele A is long ears and the trait of the recessive allele B is short ears.

    (a) What is the chance that an offspring of AA and BB parents has short ears?

    (b) What is the chance that an offspring of AB and BB parents has short ears?

12. The trait of the dominant allele A is brown hair and the trait of the recessive allele B is blond hair.

    (a) What is the chance that an offspring of AB and AB parents has brown hair?

    (b) What is the chance that an offspring of AA and AB parents has brown hair?

13. Assume that the molecule in Example 2.6 has a 25% chance of leaving the region during any 1-hour time interval. Write the dynamical system for the chance p_t that the molecule is still inside the region after t hours. After how many hours will the chance p_t fall below 10%?

14. Assume that the molecule in Example 2.6 has a 5% chance of leaving the region during any 1-hour time interval. Write the dynamical system for the chance p_t that the molecule is still inside the region after t hours. After how many hours will the chance p_t fall below 50%?

15. A molecule leaves a given region during any 1-minute interval with a 75% chance. What is the chance that the molecule will be inside the region after 2 minutes? After 3 minutes?

16. A molecule leaves a given region during any 1-minute interval with a 35% chance. What is the chance that the molecule will be inside the region after 5 minutes?

17. Consider the random walk described in Example 2.8.

    (a) What is the sample space when $t = 5$? That is, list all possible locations of a particle after completing five steps of the random walk.

    (b) What is the sample space when $t = 6$?

    (c) What is the sample space in general, after t steps of the random walk have been completed?

18. Consider the random walk described in Example 2.8.

    (a) Add two more steps to the diagram in Figure 2.7; i.e., add all possible locations of the particle for $t = 4$ and $t = 5$.

    (b) Identify the number of different ways a particle can arrive at the locations you listed in (a).

    (c) Look at the numbers in the squares in Figure 2.7 and your answer to (b). Does the pattern look familiar?

    (d) Create a table of probabilities (like Table 2.6) for the steps $t = 4$ and $t = 5$.

3	Basics of Probability Theory

We have already used "chance," "sample space," and "event," words belonging to the vocabulary of **probability theory.** In this section, we precisely define what these (and related) terms mean. Having a sound mathematical theory of probability allows us to argue correctly about the models and situations that involve **chance events.** The foundations of probability theory lie in **set theory,** which we review in this section. To find a **probability** amounts to assigning a specific number to a subset of a given set.

Sample Spaces and Events

Consider performing an experiment that involves chance (such as the experiments that we discussed in the previous two sections).

Definition 4 **Sample Space**

The *sample space* of a random experiment is the set of all possible outcomes of that experiment.

We will use S to denote a sample space.

Some experiments that we will discuss involve tossing coins and rolling dice. We assume that these objects are *fair:* tossing a fair coin means that heads (H) and tails (T) are equally likely to occur. Likewise, each number $(1, 2, 3, 4, 5,$ or $6)$ of a fair die is equally likely to be rolled.

Example 3.1 **Tossing Coins**

The possible outcomes of a single toss of a coin are heads (H) and tails (T). The sample space of the experiment that consists of tossing a fair coin once consists of two elements:

$$S = \{H, T\}$$

The outcomes of tossing a coin twice in a row are often recorded as ordered pairs, so that we know what the outcome of the first toss and what was the outcome of the second toss. For instance, heads followed by tails is recorded as (H, T); usually, we drop the parentheses and the comma and just write HT. In the experiment of tossing a coin twice, the sample space has four elements:

$$S = \{HH, HT, TH, TT\}$$

Note that HT and TH are two different events.

Example 3.2 **Rolling Dice**

The sample space of the random experiment that consists of rolling a fair die once is the set of six elements

$$S = \{1, 2, 3, 4, 5, 6\}$$

If we roll two dice simultaneously and add up the numbers that come up, the sample space describing the experiment is the set

$$S = \{2, 3, 4, 5, 6, 7, 8, 9, 10, 11, 12\}$$

Example 3.3 Sample Space for the Lion Population Stochastic Model

In our model of the lion population with immigration (Example 2.1 in Section 2), immigration is the chance factor. If we consider a single year, the sample space consists of two elements: I = "immigration of 12 lions" and N = "no immigration." If we follow the experiment for 3 years, the sample space is larger:

$$S = \{\text{III}, \text{IIN}, \text{INI}, \text{NII}, \text{NNI}, \text{NIN}, \text{INN}, \text{NNN}\}$$

For instance, INN means immigration in the first year, followed by 2 years of no immigration. (See Exercise 9.)

Example 3.4 Sample Space for AB and AB Crossing

Consider the genotype of an offspring of heterozygous parents (Example 2.3). The parents are both of genotype AB, and so the possible outcomes are AA, AB, BA, and BB. Assuming that AB and BA are the same genotype, the sample space has three elements:

$$S = \{\text{AA}, \text{AB}, \text{BB}\}$$

Example 3.5 Sample Space for the Random Walk

Recall the random walk of Example 2.8. The sample space of the experiment that consists of locating the particle at $t = 1$ (after one random move) is

$$S = \{-1, 1\}$$

When $t = 2$, the sample space of possible locations is

$$S = \{-2, 0, 2\}$$

and when $t = 3$,

$$S = \{-3, -1, 1, 3\}$$

In general, all possible locations of the particle at time t (i.e., after t random moves) is

$$S = \{-t, -t+2, \ldots, t-2, t\}$$

Note that if t is even, all numbers in S are even, and if t is odd then S is a subset of odd numbers.

Related to the outcomes of an experiment, we define the following.

Definition 5 **Simple Event and Event**

A single outcome of a random experiment is called a *simple event*. An *event* is a collection (or a set) of simple events.

Both a simple event and an event are subsets of the sample space; a simple event has one element, whereas an event can be of any size: it could be an empty set, or as large as the sample space, or anything in between.

Assume that we roll a die once. "Rolling a 3" is a simple event, and "rolling an even number," "rolling a number larger than 4," and "rolling a 6" are events.

Example 3.6 Examples of Simple Events and Events

The sample space in Example 3.3, where we investigated the immigration/no-immigration dynamics within a group of lions, is

$$S = \{\text{III}, \text{IIN}, \text{INI}, \text{NII}, \text{NNI}, \text{NIN}, \text{INN}, \text{NNN}\}$$

The event "no immigration in the first year followed by two years of immigration," or NII for short, is a simple event. The event "immigration in the third year" is not a simple event—it consists of four elements: {III, INI, NII, NNI}.

In Example 2.5 in Section 2, we analyzed the case of a dominant allele A (unattached earlobes) and a recessive allele B (attached earlobes). The sample space of the random experiment where we look at a grandchild of AA and BB parents consists of three elements:

$$S = \{AA, AB, BA\}$$

The event "offspring has attached earlobes" consists of a single element BB; thus, it is a simple event. The event "unattached earlobes" is not a simple event, since it contains two elements, AA and AB.

The sample space for the step $t = 5$ of the random walk in Example 3.5 is

$$S = \{-5, -3, -1, 1, 3, 5\}$$

The events "particle ends at -3" and "particle ends at 5" are simple events. The event "particle is more than 2 units away from the source" is not a simple event; its elements are -5, -3, 3 and 5.

Elements of Set Theory

In order to work with events, we need to learn about sets.

A *set* is defined by the elements it contains. For instance, we say "the set of all real numbers," or "the set of genotypes AA, AB, and BB," or "the set of all even numbers between 2 and 22." An *empty set* (denoted by $\emptyset$) is a set that contains no elements.

A useful way of visualizing a set is by using a *Venn diagram*. A set is represented using a geometric figure, such as a circle, an ellipse, a rectangle, or some irregular figure, with the convention that the insides of these figures contain the elements of the set. In Figure 3.1 we use Venn diagrams to represent three different sets.

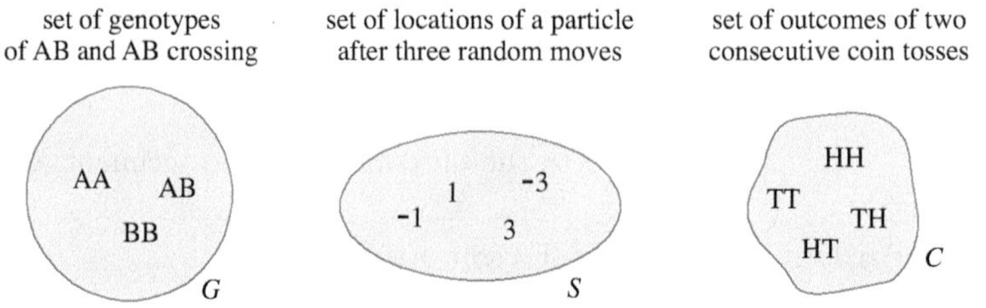

set of genotypes of AB and AB crossing

set of locations of a particle after three random moves

set of outcomes of two consecutive coin tosses

AA AB
BB
G

1 -3
-1 3
S

HH
TT TH
HT C

FIGURE 3.1

Three sets

Using set notation, we write the sets in Figure 3.1 as $G = \{AA, AB, BB\}$, $S = \{-3, -1, 1, 3\}$, and $C = \{HH, HT, TH, TT\}$.

We say that a set A is a *subset* of a set S and write $A \subseteq S$ if A is empty or contains some or all of the elements of S.

For any set A, $\emptyset \subseteq A$ and $A \subseteq A$.

We use a rectangle to visualize the sample space of an experiment and circles or ellipses to indicate its subsets. In Figure 3.2a the sample space S has two subsets, A and B.

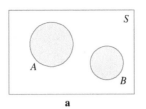 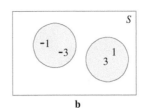

FIGURE 3.2

Set S and its subsets

Let S be the sample space $S = \{-3, -1, 1, 3\}$ of outcomes after three steps of random motion. The two subsets in Figure 3.2b contain ending locations to the right and to the left of the starting position $x = 0$.

Next, we describe the basic set operations: intersection, union, and complement.

Definition 6 Intersection of Sets

The *intersection* $A \cap B$ of sets A and B is the set of elements that belong to both A and B.

The symbol $A \cap B$ is read "A intersects B." The operation of intersection corresponds to the word "and." Thus, the intersection of events A and B consists of all simple events that belong to both A *and* B. The shaded region in Figure 3.3 represents the intersection $A \cap B$.

From the definiition, we conclude that $A \cap B = B \cap A$. Two sets A and B that have no elements in common are called *disjoint;* we write $A \cap B = \emptyset$, where $\emptyset$ is the empty set.

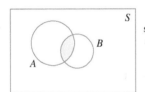

shaded region is
$A \cap B$

FIGURE 3.3

The intersection of two sets

Definition 7 Mutually Exclusive Events

Two events A and B are said to be *mutually exclusive* if they are disjoint, i.e., if $A \cap B = \emptyset$.

Consider the experiment of rolling a die once. The events "rolling an even number" and "rolling a 5" are mutually exclusive.

Regarding the immigration/no-immigration dynamics of Example 3.3: the events "immigration in the first year" $= \{\text{INN}, \text{IIN}, \text{III}, \text{INI}\}$ and "no immigration in the first 3 years" $= \{\text{NNN}\}$ are mutually exclusive.

Since $\{\text{INN}, \text{IIN}, \text{III}\} \cap \{\text{INN}, \text{NNN}\} = \{\text{INN}\} \neq \emptyset$, we conclude that the events $\{\text{INN}, \text{IIN}, \text{III}\}$ and $\{\text{INN}, \text{NNN}\}$ are not mutually exclusive.

Definition 8 Union of Sets

The *union* $A \cup B$ of sets A and B is the set that contains elements that belong to either A or B.

In other words, an element belongs to the union $A \cup B$ if it belongs to A, or to B, or to both A and B. Thus, the union corresponds to the word "or." The symbol $A \cup B$ is read "A union B."

The shaded region in Figure 3.4 represents the set $A \cup B$. The definition implies that $A \cup B = B \cup A$.

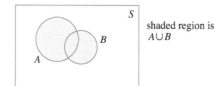

shaded region is
$A \cup B$

FIGURE 3.4

The union of two sets

Definition 9 Complement of a Set

The *complement* A^C of a subset $A \subseteq S$ is the set of all elements in S that are not in A.

See Figure 3.5. Note that the complement of a set A depends on the set that contains A (i.e., the set that A is a subset of). That set is called a *universal set*. In Definition 9, the universal set is S.

For instance, the complement of the set $A = \{1, 2\}$ as a subset of $S = \{1, 2, 3, 4\}$ is $A^C = \{3, 4\}$. The complement of A with respect to the universal set $S = \{1, 2, 3, 4, 5, 6, 7, 8, 9\}$ is $A^C = \{3, 4, 5, 6, 7, 8, 9\}$.

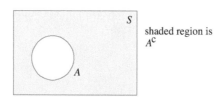

shaded region is
A^C

FIGURE 3.5

The complement of a set

In this book, the universal set S will always be the sample space of the experiment that we are investigating.

Note that $S^C = \emptyset$ and $\emptyset^C = S$. As well, $(A^C)^C = A$ for any set A. An element of S is either in A or not in A (and thus in A^C); we conclude that $A \cup A^C = S$ for any set $A \subseteq S$. No element can be in both A and A^C; thus, $A \cap A^C = \emptyset$.

Example 3.7 Operations with Events

We toss a coin three times in a row. The sample space has eight elements:

$$S = \{HHH, HHT, HTH, THH, HTT, THT, TTH, TTT\}$$

Consider the following events:

$A = $ "the outcome of the first toss was heads"

$B = $ "exactly two heads in a row occurred"

$C = $ "no more than one head occurred"

$D = $ "exactly two tails occurred"

Find $A \cup B$, $A \cap D$, $B \cap D$, D^C, and $B \cup C$. As well, express the events A^C, B^C, and C^C in words.

▶ First, we write all events as sets by listing the simple events that they contain: $A = \{HHH, HHT, HTH, HTT\}$, $B = \{HHT, THH\}$. "No more than one" means none or one, so $C = \{HTT, THT, TTH, TTT\}$. As well, $D = \{HTT, THT, TTH\}$. Thus,

$$A \cup B = \{HHH, HHT, HTH, HTT, THH\}$$

$A \cap D = \{\text{HTT}\}$ (so $A \cap D$ is a simple event)

$B \cap D = \emptyset$ (so B and D are mutually exclusive events)

$D^C = \{\text{HHH}, \text{HHT}, \text{HTH}, \text{THH}, \text{TTT}\}$

$B \cup C = \{\text{HHT}, \text{THH}, \text{HTT}, \text{THT}, \text{TTH}, \text{TTT}\}$

To express the complement of an event, we think of ways in which that event could *not* happen. Thus, we have the following:

$A^C = $ "the outcome of the first toss is not heads," or $A^C = $ "the outcome of the first toss is tails."

$B^C = $ "exactly two heads in a row did not occur."

$C^C = $ "more than one head occurred", or $C^C = $ "exactly two or exactly three heads occurred."

Example 3.8 **Operations with Events**

In the stochastic population model (see Example 1.2 in Section 1), the per capita production rate is the chance factor; the sample space is the interval $[0.95, 1.15]$. Define the events $A = $ "per capita production rate decreases the population" and $B = $ "per capita production rate increases the population."

   Using interval notation, $A = [0.95, 1)$ and $B = (1, 1.15]$. Clearly, $A \cap B = \emptyset$; i.e., A and B are mutually exclusive events. The complement $A^C = [1, 1.15]$ consists of the per capita rates that do not *decrease* the population. The value $r = 1$ (which keeps the population unchanged in number) is contained in A^C.

Note that the complement of "decreasing" is not "increasing," but "not decreasing"!

The relationship between the intersection, union, and complement is given in the following theorem.

Theorem 1 **De Morgan's Laws**

Assume that $A, B \subseteq S$. Then

(a) $(A \cap B)^C = A^C \cup B^C$

(b) $(A \cup B)^C = A^C \cap B^C$

These formulas can be verified using Venn diagrams. As an illustration, in Figure 3.6 we verify (a).

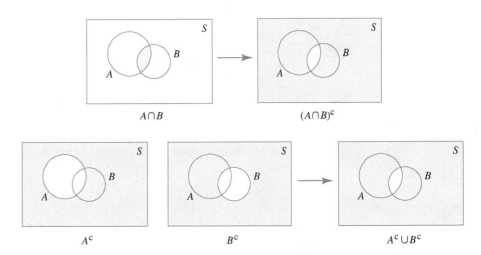

FIGURE 3.6

Verification of statement (a) from Theorem 1

Probability

A probability is a special way of assigning a number to every event in a sample space.

Definition 10   Probability

Let S denote a sample space. A *probability* is a function P that assigns, to each event A in S, a unique real number $P(A)$, called the *probability of A*. The function P satisfies the following properties:

(i) $0 \leq P(A) \leq 1$ for any event $A \subseteq S$.

(ii) $P(\emptyset) = 0$ and $P(S) = 1$.

(iii) If A and B are mutually exclusive events (i.e., $A \cap B = \emptyset$), then $P(A \cup B) = P(A) + P(B)$.

Thus, the probability is a number between 0 and 1 (and can be expressed—as is common practice—as a percent). A probability of 0 means that the event never happens, and a probability of 1 means certainty.

For instance, consider the experiment of rolling two dice and adding up the numbers that come up. If A = "sum is 1," then $P(A) = 0$, and if B = "sum is larger than 1," then $P(B) = 1$.

Recall that $A \cup A^{c} = S$ for any subset A of S. Substituting $B = A^{c}$ into the left side of (iii) in Definition 10 we get

$$P(A \cup B) = P(A \cup A^{c}) = P(S) = 1$$

The right side is

$$P(A) + P(B) = P(A) + P(A^{c})$$

(Note that we are allowed to use (iii) since A and A^{c} are mutually exclusive events.) Combining the two equations, we get $P(A) + P(A^{c}) = 1$, i.e.,

$$P(A^{c}) = 1 - P(A)$$

We have just proved the following theorem.

Theorem 2   Probability of a Complementary Event

If A is an event, then $P(A^{c}) = 1 - P(A)$.

In words, the probability that A does not happen is equal to 1 minus the probability that A happens. This sounds reasonable: if the probability that a given population reaches 1,000 bacteria is 70%, then the probability that it will not reach 1,000 is $100\% - 70\% = 30\%$.

Note that (iii) in Definition 10 requires that A and B be disjoint events (mutually exclusive). If the probability that in a random experiment a monkey presses a red button is 0.2 and the probability that it presses a green button is 0.3, and there is no chance that it can press both buttons at the same time, then the probability that the monkey presses either a red button or a green button is $0.2 + 0.3 = 0.5$.

In the case where two events are not mutually exclusive, we use the following result.

Theorem 3 Probability of the Union of Two Events

If A and B are two events, then

$$P(A \cup B) = P(A) + P(B) - P(A \cap B) \qquad (3.1)$$

In the special case where A and B are mutually exclusive events, (3.1) reduces to $P(A \cup B) = P(A) + P(B)$, which is the requirement (iii) from Definition 10.

Intuitively, (3.1) is clear: the intersection $A \cap B$ is contained in both A and B. So in calculating $P(A) + P(B)$ we have counted the probability of $A \cap B$ twice and need to subtract one of these contributions. The formal proof of this formula is not difficult. We discuss it in Exercise 39.

We are ready to answer one of the questions we asked earlier: assume that there is a 60% chance that it will rain on Saturday and a 50% chance that it will rain on Sunday. What is the chance that it will rain on the weekend?

Define the events $A =$ "it will rain on Saturday" and $B =$ "it will rain on Sunday." Then $A \cup B =$ "it will rain on the weekend." We are given that $P(A) = 0.6$ and $P(B) = 0.5$ and are asked to calculate $P(A \cup B)$.

Clearly, $P(A) + P(B) = 0.6 + 0.5 = 1.1 > 1$ cannot be the probability of any event. Looking at (3.1), we see that in order to compute $P(A \cup B)$, we need to know $P(A \cap B)$, i.e., the probability that it will rain on *both* Saturday and Sunday. If we know that information (say, $P(A \cap B) = 0.4$), then the probability that it will rain on the weekend is

$$P(A \cup B) = P(A) + P(B) - P(A \cap B) = 0.6 + 0.5 - 0.4 = 0.7$$

i.e., 70%.

The probability can be (and is) interpreted as an *area* upon adopting the following conventions: to satisfy (ii) in Definition 10, we take the area of the sample space S to be 1 and the area of the empty set to be 0. Any other event (subset of S) has area between 0 and 1 (that's (i) from Definition 10). Thus, for $A \subseteq S$,

$$P(A) = \text{area of } A \qquad (3.2)$$

The following arguments will convince us that this makes a lot of sense.

(a) If A and B are disjoint, then the area of $A \cup B$ is the sum of the area of A and the area of B, which is exactly what (iii) in Definition 10 requires.

(b) If the area of A is $P(A)$, then

$$\text{area of } A^{\mathrm{c}} = \text{area of } S - \text{area of } A = 1 - P(A)$$

i.e., $P(A^{\mathrm{c}}) = 1 - P(A)$.

(c) If A and B are not disjoint, then the area of $A \cup B$ cannot be equal to the sum of the areas of A and B, since the region $A \cap B$ has been counted in twice. Thus,

$$\text{area of } (A \cup B) = \text{area of } A + \text{area of } B - \text{area of } (A \cap B)$$

which is exactly what we claim in (3.1).

Assigning Probability to Events

Now that we know what the probability is, we need to work on a major question —how to assign probabilities to events within a given sample space S so that all requirements of Definition 10 are fulfilled.

The approaches to answering this question differ based on the size of S. We first focus on *finite* sample spaces. Later, we learn how to do it for *infinite* sample spaces.

Assume that the sample space of an experiment is finite, i.e., that it contains n distinct, simple events $E_1, E_2, \ldots, E_n$. By virtue of being simple and distinct, the events $E_1, E_2, \ldots, E_n$ are all mutually exclusive. As well,

$$S = E_1 \cup E_2 \cup \cdots \cup E_n \tag{3.3}$$

To each event E_i, $i = 1, 2, \ldots, n$, we assign the number $P(E_i)$ between 0 and 1. From (3.3) we conclude that

$$P(S) = P(E_1) + P(E_2) + \cdots + P(E_n) \tag{3.4}$$

Since $P(S) = 1$, when assigning probabilities $P(E_i)$ we have to make sure that they add up to 1. (The formula in (iii) from Definition 10 extends to a finite union of mutually exclusive events (see Exercise 40) and that's why (3.4) follows from (3.3).)

Example 3.9   Assigning Probabilities

The sample space of the experiment that consists of tossing a fair coin once is $S = \{H, T\}$. To the two simple events $\{H\}$ and $\{T\}$ we assign $P(H) = 0.5$ and $P(T) = 0.5$.

In the model of a population of lions with immigration (Example 2.1), the chance factor is immigration. Defining $A =$ "immigration occurs," we let $P(A) = 0.5$. Its complement $A^C =$ "immigration does not occur" has the same probability $P(A^C) = 0.5$. The sample space consists of two simple events, A and A^C.

Assume that, at this moment, a virus is present in a population. The chance that it will be present the following month is 75%. To model this situation, we define the events $A =$ "virus is present the following month" and $B = A^C =$ "virus is not present the following month." We assign $P(A) = 0.75$ and $P(B) = 0.25$. ◣

Example 3.10   Assigning Probabilities: Random Walk

Consider the second step in the random walk (discussed in detail in Example 2.8 in Section 2), shown in Figure 3.7.

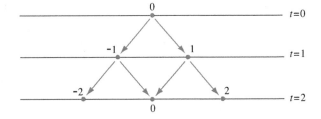

FIGURE 3.7

Second step in the random walk

Recall that, in each step, a particle moves either left or right for 1 unit with equal probability. The outcomes of the random walk are the locations L of the particle after two steps, and so the sample space is $S = \{-2, 0, 2\}$.

There are three simple events, $E_1 = \{-2\}$, $E_2 = \{0\}$, and $E_3 = \{2\}$. Looking at the diagram, we see that of the total of four paths, one each ends at -2 and 2, whereas two paths end at 0 (see Figures 2.6 or 2.7 in Section 2).

In Table 3.1 we assign the probabilities that the particle ends at a location L. Clearly, the probabilities add up to 1.

The events $E_1 = \{L = -2\}$ and $E_2 = \{L = 0\}$ are mutually exclusive. Thus,

$$P(L = -2 \text{ or } L = 0) = P(E_1 \cup E_2)$$
$$= P(E_1) + P(E_2)$$
$$= P(L = -2) + P(L = 0) = 3/4$$

In words, there is a 75% chance that the particle ends up at location -2 or at location 0.

Table 3.1

Event	Probability
$P(L = -2)$	$1/4$
$P(L = 0)$	$1/2$
$P(L = 2)$	$1/4$

There is one situation where assigning probabilities is straightforward. Assume that a sample space S consists of n $(n \geq 1)$ simple events $E_1, E_2, \ldots, E_n$ that are *equally likely*, i.e., $P(E_1) = P(E_2) = \cdots = P(E_n)$. From

$$P(E_1) + P(E_2) + \cdots + P(E_n) = 1$$

we conclude that $P(E_i) = 1/n$ for all i.

Any event A can be written as the union of a certain number (say, k) of simple events. Thus,

$$P(A) = \underbrace{\frac{1}{n} + \frac{1}{n} + \cdots + \frac{1}{n}}_{k} = \frac{k}{n}$$

Keep in mind that n is the number of elements in S.

For a given set X, we use $|X|$ to denote the number of elements in X. We summarize our discussion in the statement of the following theorem.

Theorem 4 **Assigning Probabilities: Equally Likely Simple Events**

Assume that S is a finite sample space in which all outcomes (simple events) are equally likely. The probability of an event $A \subseteq S$ is

$$P(A) = \frac{|A|}{|S|}$$

Example 3.11 Assigning Probabilities: Three Tosses of a Coin

The sample space of the experiment that consists of tossing a coin three times in a row

$$S = \{\text{HHH}, \text{HHT}, \text{HTH}, \text{HTT}, \text{THH}, \text{THT}, \text{TTH}, \text{TTT}\}$$

contains eight equally likely simple events. In Example 3.7 we defined four events

$A =$ "the outcome of the first toss was heads" $= \{\text{HHH}, \text{HHT}, \text{HTH}, \text{HTT}\}$

$B =$ "exactly two heads in a row occurred"$= \{\text{HHT}, \text{THH}\}$

$C =$ "no more than one head occurred"$= \{\text{HTT}, \text{THT}, \text{TTH}, \text{TTT}\}$

$D =$ "exactly two tails occurred"$= \{\text{HTT}, \text{THT}, \text{TTH}\}$

With Theorem 4 in mind, we conclude that $P(A) = 4/8 = 1/2$, $P(B) = 2/8 = 1/4$, $P(C) = 4/8 = 1/2$, and $P(D) = 3/8$.

Example 3.12 Assigning Probabilities: Rolling Two Dice

The experiment consists of rolling two dice, and the outcome is the sum of the numbers that come up. The space of all possible outcomes consists of 36 equally likely events

$$S = \{(1,1), (1,2), (1,3), (1,4), (1,5), (1,6), (2,1), \ldots, (6,5), (6,6)\}$$

where the ordered pair (m, n) records that the number m came up on the first die and n came up on the second $(1 \leq m, n \leq 6)$. Since all events are equally likely (we assume that the dice are fair), the probability that any one occurs is $1/36$.

The probabilities of the events "sum $= 2$," "sum $= 3$," ..., "sum $= 12$" are given in Table 3.2. They were computed using Theorem 4. For instance, the event "sum $= 6$" consists of five simple events, $\{(1, 5), (2, 4), (3, 3), (4, 2), (5, 1)\}$. Therefore,

$$P(\text{sum} = 6) = \frac{|\{(1, 5), (2, 4), (3, 3), (4, 2), (5, 1)\}|}{36} = \frac{5}{36}$$

Based on Table 3.2 we can calculate the probabilities of other events. For instance, if $A =$ "sum is larger than 9," then (using mutual exclusivity) we get

$$P(A) = P(\text{sum} = 10) + P(\text{sum} = 11) + P(\text{sum} = 12)$$
$$= \frac{3}{36} + \frac{2}{36} + \frac{1}{36} = \frac{6}{36} = \frac{1}{6}$$

The probability that we roll a sum divisible by 5 is (call the event B)

$$P(B) = P(\text{sum} = 5) + P(\text{sum} = 10) = \frac{4}{36} + \frac{3}{36} = \frac{7}{36}$$

Table 3.2

Simple events	Sum	Probability
(1,1)	2	1/36
(1,2), (2,1)	3	2/36
(1,3), (2,2), (3,1)	4	3/36
(1,4), (2,3), (3,2), (4,1)	5	4/36
(1,5), (2,4), (3,3),(4,2),(5,1)	6	5/36
(1,6), (2,5), (3,4), (4,3), (5,2), (6,1)	7	6/36
(2,6), (3,5), (4,4), (5,3), (6,2)	8	5/36
(3,6), (4,5), (5,4), (6,3)	9	4/36
(4,6), (5,5), (6,4)	10	3/36
(5,6), (6,5)	11	2/36
(6,6)	12	1/36

Example 3.13   Assigning Probabilities: Genotypes

Suppose that both parents are of genotype AB (for a background on the genetics needed, see Example 2.3 in Section 2). Their offspring will have one of the four combinations of alleles: AA, AB, BA, and BB. The sample space of genotypes has three elements, since AB and BA are the same genotype. The probabilities are given in Table 3.3.

Define the events $X =$ "offspring has at least one A allele" and $Y =$ "offspring is homozygous." Their probabilities are

$$P(X) = P(\text{AA}) + P(\text{AB}) = \frac{1}{4} + \frac{1}{2} = \frac{3}{4}$$

and

$$P(Y) = P(\text{AA}) + P(\text{BB}) = \frac{1}{4} + \frac{1}{4} = \frac{1}{2}$$

Table 3.3

Genotype	Probability
AA	1/4
AB	1/2
BB	1/4

Remarks   Note that in applications it is the context that determines the probabilities. Mathematically, assigning $P(AA) = 0.8$, $P(AB) = 0.05$, and $P(BB) = 0.15$ in Example 3.13 is sound, but biologically it might not make sense. In other words, there are many (infinitely many) different ways of assigning probabilities to events, but very few might make sense biologically (or within the context of some other discipline).

A good way to verbalize information about chance is to use the *relative frequencies* instead of reporting the probability of an event occurring. The relative frequency is a fraction a/b, usually referred to as "a out of b."

Consider an example. The chance that an adult male in Canada will have a heart attack is 0.0153 (this is probability language). We can translate the decimal number into 1.53% (still probability language). We can rephrase the given fact as follows: 1.53 out of 100 (relative frequency language) adult males will have a heart attack. To avoid decimal numbers, we multiply by 100: 153 of 10,000 (relative frequency language) adult males will have a heart attack.

Definition 11   Odds

The *odds in favour of an event* are represented as the ratio of integers a/b, which is equal to the quotient of the probability that the event will occur and the probability that the event will not occur.

Quite often, $a : b$ is used to express the odds a/b. Instead of saying "the odds in favour of an event" we say "the odds for." Thus, if p is the probability that an event A occurs, then the odds for A are $p/(1 - p)$ or $p : (1 - p)$.

For instance, the probability that we roll a 3 in a single roll of a die is $1/6$. Thus, the odds for a 3 are

$$\frac{p}{1 - p} = \frac{1/6}{1 - 1/6} = \frac{1/6}{5/6} = \frac{1}{5}$$

or $1 : 5$.

Unlike probability, odds can be greater than 1. Assume that we toss a coin twice, and define the event $A =$ "at least one H is tossed." Then $P(A) = P(\{TH, HT, HH\}) = 3/4$, and the odds in favour of A are

$$\frac{p}{1 - p} = \frac{3/4}{1 - 3/4} = \frac{3}{1}$$

i.e., three to one $(3 : 1)$.

Summary   The **sample space** of an experiment is the set of all possible outcomes. A single outcome is called a **simple event,** and any collection of outcomes (i.e., any subset of the sample space) defines an **event.** By using set operations—union, intersection, and complement—we form new events. The **probability** is a function that assigns a number between 0 and 1 to every event in a sample space. The empty set has probability 0, and the sample space has probability 1. The probability is **additive** for **mutually exclusive events:** if two events have nothing in common, then the probability of their union is equal to the sum of the probabilities of the events. To **calculate probabilities,** we start by assigning probabilities to all simple events in the sample space; then, using their mutual exclusivity, we can find the probability of any event in the sample space. If a sample space consists of **equally likely simple events,** then the probability of an event occurring is the number of simple events in that event divided by the number of simple events in the sample space.

3	Exercises

1. Describe an experiment whose sample space consists of three simple events that are not equally likely. Give an example of an experiment whose sample space consists of three equally likely simple events.

2. Describe an experiment whose sample space consists of ten simple events that are not equally likely. Give an example of an experiment whose sample space consists of ten equally likely simple events.

▽ 3–6 ▪ List all elements of the sample space S for each experiment. What is $|S|$ (i.e., the number of elements in S)?

3. We roll two dice simultaneously and multiply the numbers that come up.

4. We toss a fair coin ten times and calculate the following difference: the number of heads minus the number of tails.

5. We roll two dice one after the other, and as we repeat this routine eight times, we count the number of times the first number is larger than the second.

6. We toss a fair coin four times.

7. We toss a coin n times, $n \geq 1$. How many elements does the sample space have?

8. We roll a die n times, $n \geq 1$. What is the size of the sample space?

9. Continue with Example 3.3. How many elements does the sample space have if we consider four and five years of the immigration/no-immigration dynamics? List all of them. Consider the immigration/no-immigration dynamics for n years. How many elements does the sample set have?

10. In order to prove that two events A and B are mutually exclusive, we need to show that $A \cap B = \emptyset$. To prove that A, B, and C are mutually exclusive, we need to prove that $A \cap B = \emptyset$, $A \cap C = \emptyset$, and $B \cap C = \emptyset$.

   (a) List all conditions that we need to check in order to prove that the four events A, B, C, and D are mutually exclusive.

   (b) How many conditions do we need to check to prove that n $(n \geq 1)$ events are mutually exclusive?

▽ 11–14 ▪ For the given universal set S and the subsets A and B, find $A \cup B$, $A \cap B$, A^{c}, and $A \cap B^{\mathrm{c}}$.

11. $S = \{1, 2, 3, 4, 5, 6, 7, 8, 9\}$; $A = \{1, 3, 5, 6, 7, 8, 9\}$, $B = \{1, 3, 4\}$

12. $S = \{a, b, c, d, e, f, g\}$; $A = \{a, f\}$, $B = \{c, d, e\}$

13. S is the set of all non-negative integers, $S = \{0, 1, 2, 3, \ldots\}$; A is the set of even numbers (take 0 to be even) and B is the set of all numbers divisible by 4.

14. S is the set of all non-negative integers, $S = \{0, 1, 2, 3, \ldots\}$; A is the set of even numbers (take 0 to be even) and B is the set of all numbers divisible by 3.

15. Use Venn diagrams to show that part (b) of Theorem 1 is true, i.e., $(A \cup B)^{\mathrm{c}} = A^{\mathrm{c}} \cap B^{\mathrm{c}}$.

16. Given that $P(A \cap B) = 0.2$ and $P(A \cap B^{\mathrm{c}}) = 0.45$, find $P(A)$.

17. Given that $P(A) = 0.4$, $P(B) = 0.2$, and $P(A \cap B) = 0.1$, find $P(A^{\mathrm{c}} \cap B^{\mathrm{c}})$.

18. Given that $P(A) = 0.3$, $P(B) = 0.2$, and $P(A \cup B) = 0.4$, find $P(A \cap B)$ and $P(A^{\mathrm{c}} \cap B^{\mathrm{c}})$.

19. Assume that A is a subset of B and $A \neq B$ (i.e., A is a *proper subset* of B). Show that $P(A) < P(B)$. [Hint: Using Venn diagrams, show that $B = A \cup (A^{\mathrm{c}} \cap B)$.]

20. Explain why it is not possible to assign probabilities to A and B in the following way: $P(A) = 0.1$, $P(B) = 0.2$, and $P(A \cap B) = 0.4$.

21. Explain why it is not possible to assign probabilities to A and B in the following way: $P(A) = 0.5$, $P(B) = 0.2$, and $P(A \cap B) = 0.4$.

22. Explain why it is not possible to assign probabilities to A and B in the following way: $P(A) = 0.1$, $P(B) = 0.2$, and $P(A \cup B) = 0.4$.

▽ 23–26 ▪ Given are the sample set S and the assignment of probabilities for all but one simple event. Find the requested probabilities and answer the questions.

23. $S = \{1, 2, 3, 4, 5\}$; $P(1) = 0.4$, $P(2) = 0.15$, $P(3) = 0.2$, $P(5) = 0.1$

   (a) Find $P(4)$.

   (b) Let $A = \{1, 2\}$ and $B = \{2, 3, 4\}$. Find $P(A)$, $P(B)$, and $P(A \cup B)$.

   (c) Is $P(A \cup B)$ equal to $P(A) + P(B)$? Why or why not?

24. $S = \{1, 2, 3, 4, 5\}$; $P(1) = 0.1$, $P(2) = 0.1$, $P(4) = 0.1$, $P(5) = 0.2$

   (a) Find $P(3)$.

   (b) Let $A = \{1, 2, 3\}$ and $B = \{4, 5\}$. Find $P(A)$, $P(A^C)$, $P(B)$, and $P(A \cup B)$.

   (c) Is $P(A \cup B)$ equal to $P(A) + P(B)$? Why or why not?

25. $S = \{1, 2, 3, 4, 5\}$; $P(1) = 0.2$, $P(3) = 0.4$, $P(4) = 0.3$, $P(5) = 0.1$

   (a) Find $P(2)$.

   (b) Let $A = \{2\}$ and $B = \{1, 3, 4, 5\}$. Find $P(A)$, $P(A^C)$, $P(B)$, and $P(B^C)$.

   (c) Let C be any event. Is $P(A \cup C)$ equal to $P(A) + P(C)$? Why or why not?

26. $S = \{1, 2, 3, 4, 5\}$; $P(1) = 0.2$, $P(2) = 0.2$, $P(4) = 0.2$, $P(5) = 0.2$

   (a) Find $P(3)$.

   (b) What is the probability of any event that consists of four simple events?

   (c) Assume that A and B consist of three simple events each. Is $P(A \cup B)$ equal to $P(A) + P(B)$? Why or why not?

27. (a) We toss a fair coin three times. What is the probability of getting exactly two heads in a row?

   (b) We toss a fair coin four times. What is the probability of getting exactly two heads in a row?

28. We roll two fair dice. What is the probability that the sum of the numbers that come up is odd?

29. We roll two fair dice. What is the probability that the maximum of the two numbers is 4?

▽ 30–33 ▪ Assume that female and male children are equally likely to be born.

30. A family has three children. Find the probability that all three are girls.

31. A family has three children. Find the probability that at least one child is a girl.

32. A family has five children. Find the probability that all five are girls.

◣ 33. A family has six children. Find the probability that at least one child is a girl.

34. The odds for A are $2 : 7$. What is the probability of A occurring?

35. The odds for an event are $2 : 100$. What is the probability of the event occurring?

36. If the probability of an event A occurring is 0.4, what are the odds for A?

37. Consider the following modification of the random walk routine: a particle is released from $x = 0$ at $t = 0$; during each time interval, with a probability of 1/3, it moves left for 1 unit, or right for 1 unit, or stays where it is.

    (a) Find the sample space S when $t = 1$ (i.e., after one step of random motion), and assign probabilities to each simple event in S.

    (b) Find the sample space S when $t = 2$ (i.e., after two steps of random motion), and assign probabilities to each simple event in S.

    (c) Find the sample space when $t = 3$, and assign probabilities to each simple event.

38. Consider the following modification of the random walk routine: a particle is released from $x = 0$ at $t = 0$; during each time interval, with a probability of 1/2, it moves left for 1 unit, or right for 2 units.

    (a) Find the sample space S when $t = 2$ (i.e., after two steps of random motion), and assign probabilities to each simple event in S.

    (b) Find the sample space when $t = 3$ and $t = 4$ and assign probabilities to each simple event.

39. We prove that $P(A \cup B) = P(A) + P(B) - P(A \cap B)$ for two sets A and B. The idea is to write $A \cup B$ and A as disjoint unions.

    (a) Using Venn diagrams, show that $A \cup B$ is a disjoint union of B and $B^C \cap A$.

    (b) Using Venn diagrams, show that A is a disjoint union of $A \cap B$ and $A \cap B^C$.

    (c) Apply (iii) from Definition 10 to (a) and (b) and combine the two equations that you obtain.

40. Show that, for three mutually exclusive events A, B, and C, $P(A \cup B \cup C) = P(A) + P(B) + P(C)$. Explain how you would prove the general claim: if $E_1, E_2, \ldots, E_n$ are mutually exclusive events, then $P(S) = P(E_1) + P(E_2) + \cdots + P(E_n)$, where $S = E_1 \cup E_2 \cup \cdots \cup E_n$, and $n \geq 2$. [Hint: Write $A \cup B \cup C = A \cup (B \cup C)$ and use (iii) from Definition 10.]

4	Conditional Probability and the Law of Total Probability

The probability we defined in the previous section is also referred to as the **unconditional probability:** we calculate the chance of an event occurring disregarding any factors that might affect it. However, there are situations where we need to calculate the probability of an event occurring knowing (or assuming) that another event has taken place. That's the motivation behind introducing **conditional probability.**

Conditional Probability

We start with an example.

Example 4.1   Green-Eyed and Red-Eyed Kittens

Consider a population of cats with the following genetics. The trait of the dominant allele R is red eyes; the trait of the recessive allele G is green eyes. This means that there are three genotypes (RR, RG, and GG) and two phenotypes: red eyes (RR, RG) and green eyes (GG).

The sample space of genotypes of a kitten both of whose parents are RG is

$$S = \{RR, RG, GG\}$$

and the probabilities are (see Example 3.13 in Section 3)

$$P(RR) = P(GG) = 1/4, \quad P(RG) = 1/2 \qquad (4.1)$$

We ask the following question: RG parents have a kitten. We look at the kitten and see that it has red eyes. What is the probability that its genotype is RR?

Note the important difference—we are not asking the question "What is the probability that a kitten has the combination RR?" (we know that it's 1/4); instead, our question is "*Given* that the kitten has red eyes (i.e., under the *condition* that the kitten has red eyes), what is the probability that its genotype is RR?"

In a moment we will introduce the formula that will help us calculate this probability. But first we argue intuitively, using relative frequencies.

Take a sample of 1,000 kittens, all from RG parents. Based on (4.1), ideally, 250 kittens have genotype RR, 250 kittens have genotype GG, and 500 kittens have genotype RG; see Figure 4.1a. The sample space consists of all 1,000 kittens.

We know that the kitten has red eyes. This eliminates the group of 250 GG kittens. We draw a new diagram, this time with the sample space consisting of 750 red-eyed kittens; Figure 4.1b. Of 750 red-eyed kittens, 250 are of genotype RR. Thus, the probability that a red-eyed kitten has genotype RR is 250/750 = 1/3. Note that this probability is not equal to 1/4, which is the probability that the kitten is RR, based on the sample space of all 1,000 kittens.

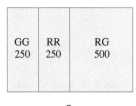
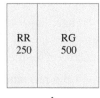

FIGURE 4.1

Reduced sample space

The condition imposed (red eyes) reduces the size of the sample space, and we calculate the probability based on that *reduced* sample space.

To denote conditional probability, we use

$$P(\text{event} \mid \text{condition})$$

In our case,

$$P(\text{RR} \mid \text{red eyes}) = \frac{1}{3}$$

Formally, if $A =$ "genotype RR" and $C =$ "red eyes," then $P(A \mid C) = 1/3$. ◣

Definition 12   Conditional Probability

The probability of an event A *conditional* on an event C is given by

$$P(A \mid C) = \frac{P(A \cap C)}{P(C)}$$

provided that $P(C) \neq \emptyset$. ◣

Instead of saying "the probability of A conditional on C" we can say "the probability of A, given C."

Example 4.2   **Example 4.1, Continued**

Using the probabilities in (4.1), we get

$P(A) = P(\text{genotype RR}) = 1/4$

$P(C) = P(\text{red eyes}) = P(\text{RR}, \text{RG}) = 1/4 + 1/2 = 3/4$

$P(A \cap C) = P(\text{genotype RR and red eyes}) = P(\text{RR}) = 1/4$, since "genotype RR" guarantees (is a subset of) "red eyes." Therefore,

$$P(A \mid C) = \frac{P(A \cap C)}{P(C)} = \frac{1/4}{3/4} = \frac{1}{3}$$

which confirms our reasoning in Example 4.1. ◣

Note that the conditional probability does not "commute"; in general, $P(A \mid C) \neq P(C \mid A)$, as the following example shows.

Example 4.3   **Computing Conditional Probability: Red-Eyed and Green-Eyed Kittens**

Keep the context of Example 4.1. Define $B =$ "kitten is of homozygous type" and $C =$ "kitten has red eyes." Find and interpret $P(B \mid C)$ and $P(C \mid B)$.

▶ By Definition 12,

$$P(B \mid C) = \frac{P(B \cap C)}{P(C)}$$

We have already calculated that $P(C) = 3/4$. Using the probabilities given in (4.1), we compute

$$P(B \cap C) = P(\text{homozygous and red eyes}) = P(\text{RR}) = 1/4$$

Thus,

$$P(B \mid C) = \frac{P(B \cap C)}{P(C)} = \frac{1/4}{3/4} = \frac{1}{3}$$

is the probability that a kitten is homozygous, given that it has red eyes. Since

$$P(B) = P(\text{RR or GG}) = P(\text{RR}) + P(\text{GG}) = \frac{1}{2}$$

it follows that

$$P(C \mid B) = \frac{P(C \cap B)}{P(B)} = \frac{1/4}{1/2} = \frac{1}{2}$$

Thus, there is a fifty-fifty chance that a kitten has red eyes, given that it is of homozygous genotype.

Example 4.4 **Computing Conditional Probability: Incidence of Heart Attacks in Canada**

Based on a survey that examined the 1950–1999 data on cardiovascular diseases in Canada, 1.53% of adult Canadians who suffered a heart attack were male, and 0.54% of adult Canadians who suffered a heart attack were female [Source: Manuel, D., Leung, M., Nguyen, K., Tanuseputro, P., & Johansen, H. (2003). Burden of cardiovascular disease in Canada. *Canadian Journal of Cardiology*, 19 (9), 997-1004.]

To organize the information (whenever probability is involved), we define events. Let

$F =$ "person is female"

$M =$ "person is male"

$H =$ "person had a heart attack"

We assume that in Canada, $P(F) = P(M) = 0.5$. From the given information, we know that

$P(M \cap H) = P(\text{male and had a heart attack}) = 0.0153$

$P(F \cap H) = P(\text{female and had a heart attack}) = 0.0054$

Answer the following questions:

(a) What is the probability that an adult male had a heart attack? That is, what is $P(\text{heart attack} \mid \text{male})$?

(b) What is the probability that a randomly chosen adult Canadian had a heart attack?

(c) What is the probability that a person who had a heart attack is male?

▶ (a) By Definition 12,

$$P(H \mid M) = \frac{P(H \cap M)}{P(M)} = \frac{0.0153}{0.5} = 0.0306$$

or 3.06%.

(b) We are asked to find $P(H)$. Looking at Figure 4.2, we see that

$$H = (H \cap M) \cup (H \cap F)$$

(i.e., a person who had a heart attack is either male or female).

FIGURE 4.2

Using Venn diagrams to show that
$H = (H \cap M) \cup (H \cap F)$

$H =$ shaded region

Since $H \cap M$ and $H \cap F$ are mutually exclusive events, the probability of their union is the sum of the probabilities; thus,

$$P(H) = P((H \cap M) \cup (H \cap F))$$
$$= P(H \cap M) + P(H \cap F) = 0.0153 + 0.0054 = 0.0207$$

(c) We know that a person had a heart attack; thus, "heart attack" is the condition, and we need to find $P(\text{male}\,|\,\text{heart attack})$, i.e., $P(M\,|\,H)$. We compute

$$P(M\,|\,H) = \frac{P(M \cap H)}{P(H)} = \frac{0.0153}{0.0207} \approx 0.739$$

or about 74%.

Example 4.5   **Using Relative Frequences to Argue about Probabilities in Example 4.4**

Take a sample of 20,000 Canadians, 10,000 females and 10,000 males. The fraction of the whole population (out of 20,000) who are males and had a heart attack is 0.0153. Thus, in the group of 20,000 Canadians, $(0.0153)(20,000) = 306$ males had a heart attack. Likewise, $(0.0054)(20,000) = 108$ females had a heart attack; see Figure 4.3a.

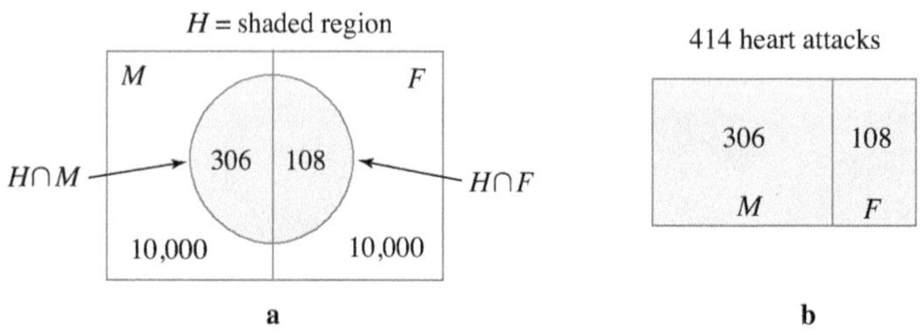

FIGURE 4.3

Heart attacks: relative frequencies

Of 10,000 males, 306 had a heart attack; thus, the probability that a male had a heart attack is $306/10{,}000 = 0.0306$ (this answers question (a) from Example 4.4).

In the total population of 20,000 Canadians, $306 + 108 = 414$ people had a heart attack; Figure 4.3b. Knowing that a heart attack occurred, the probability that it affected a male is $306/414 \approx 0.739$ (this is the answer to (c) in Example 4.4).

From Figure 4.3a we see that of 10,000 females, 108 experienced a heart attack. Thus, the probability that a female Canadian experienced a heart attack is $108/10{,}000 = 0.0108$. Checking using conditional probability, we get

$$P(H\,|\,F) = \frac{P(H \cap F)}{P(F)} = \frac{0.0054}{0.5} = 0.0108$$

Of the 414 Canadians who had a heart attack, 108 were females. Thus, the probabilit that a person who had a heart attack is a female is $108/414 \approx 0.261$.

There are alternatives to the calculation we have just finished: using complementary events, we find that

$$P(\text{person who had a heart attack is female})$$
$$= 1 - P(\text{person who had a heart attack is male})$$
$$\approx 1 - 0.739 = 0.261$$

Another way to do the same is to use the conditional probability

$$P(F\,|\,H) = \frac{P(F \cap H)}{P(H)} = \frac{0.0054}{0.0207} \approx 0.261$$

To deepen our understanding of conditional probability, we look into its properties and work through a few more examples.

From the formulas for the conditional probability

$$P(A\,|\,C) = \frac{P(A \cap C)}{P(C)} \quad \text{and} \quad P(C\,|\,A) = \frac{P(C \cap A)}{P(A)}$$

we get

$$P(A \cap C) = P(A \mid C)P(C) \tag{4.2}$$

and

$$P(C \cap A) = P(C \mid A)P(A) \tag{4.3}$$

Thus, the probability of $A \cap C$ can be calculated in two different ways.

To illustrate these formulas, we go back to the green-eyed and red-eyed kittens from the begining of this section. Recall that in Example 4.2 we defined the events $A =$ "kitten is of genotype RR" and $C =$ "kitten has red eyes" and showed that $P(A) = 1/4$, $P(C) = 3/4$, and $P(A \mid C) = 1/3$.

We can compute

$$P(\text{genotype RR and red eyes}) = P(A \cap C)$$

in two different ways. A kitten can have red eyes and have genotype RR conditional on red eyes; thus

$$P(\text{genotype RR and red eyes}) = P(\text{genotype RR} \mid \text{red eyes})P(\text{red eyes})$$
$$= P(A \mid C)P(C)$$
$$= \frac{1}{3} \cdot \frac{3}{4} = \frac{1}{4}$$

Or, a kitten can be of genotype RR and have red eyes conditional on genotype RR; thus

$$P(\text{genotype RR and red eyes}) = P(\text{red eyes} \mid \text{genotype RR})P(\text{genotype RR})$$
$$= P(C \mid A)P(A)$$
$$= 1 \cdot \frac{1}{4} = \frac{1}{4}$$

(Note that $P(\text{red eyes} \mid \text{genotype RR}) = 1$ since, given RR, it is guaranteed that the kitten will have red eyes.)

We consider $P(A \mid C)$ in two extreme cases.

If A and C are mutually exclusive ($A \cap C = \emptyset$), then

$$P(A \mid C) = \frac{P(A \cap C)}{P(C)} = \frac{P(\emptyset)}{P(C)} = \frac{0}{P(C)} = 0$$

This sounds right. We know that C occurred, and so we are sure that A could not have occurred.

If A is a subset of C, then

$$P(C \mid A) = \frac{P(C \cap A)}{P(A)} = \frac{P(A)}{P(A)} = 1$$

Again, it makes sense. We know that A occurred, and since A is within C, this means that C occurred as well.

Example 4.6   **Red-Eyed and Green-Eyed Kittens, Again**

The events "green eyes" and "genotype RR" are mutually exclusive. Thus,

$$P(\text{genotype RR} \mid \text{green eyes}) = 0$$

and

$$P(\text{green eyes} \mid \text{genotype RR}) = 0$$

The event "genotype RR" is a subset of the event "red eyes," and we conclude that

$$P(\text{red eyes} \mid \text{genotype RR}) = 1$$

Example 4.7   Conditional Probability with Apples and Bananas

The owners of farm X grow apples. The owners of farm Y grow apples on one half of their farm and bananas on the other half. Define the events X = "farm X," Y = "farm Y," A = "apples," and B = "bananas."

The conditional probability $P(X\,|\,B)$ is zero, since X and B are mutually exclusive. In other words, if we arrive at one of the two farms and see bananas, then we are sure that it is not farm X.

To find $P(Y\,|\,B)$ means to answer the following question: given that we see bananas, what is the probability that we are on farm Y? Since bananas grow only on farm Y, we conclude that $P(Y\,|\,B) = 1$. (Note that this is the case of the probability of an event conditional on its subset.)

The probability $P(B\,|\,X) = 0$, because B and X are mutually exclusive. Alternatively: given that we are on farm X, it is certain that we will see no bananas there.

Now assume that we are on farm Y. We might be standing among apples or among bananas. Thus, $P(B\,|\,Y)$ is a number smaller than 1. It is given that bananas grow on one half of farm Y, and so $P(B\,|\,Y) = 1/2$.

Note that $P(B\,|\,Y)$ is larger than $P(B)$; for $P(B\,|\,Y)$, the sample set is farm Y, and for $P(B)$ the sample set consists of the two farms combined. We can check this reasoning algebraically:

$$P(B\,|\,Y) = \frac{P(B \cap Y)}{P(Y)} = \frac{P(B)}{P(Y)} > P(B)$$

B is a subset of Y, and thus $B \cap Y = B$ and $P(B \cap Y) = P(B)$. In the last step, the inequality holds because we are dividing $P(B)$ by the number $P(Y)$, which is smaller than 1.   ◣

The Law of Total Probability

In Example 4.4, we studied the incidence of heart attacks among adult Canadians. We divided the total population (call it S) into two events (subsets): females (F) and males (M).

The events M and F are *mutually exclusive* (i.e., $M \cap F = \emptyset$) and *collectively exhaustive;* that is, their union is equal to all of the sample space S, $M \cup F = S$. We say that M and F form a *partition* of S.

Take an event $H \neq \emptyset$. We can write H as the union (see Figure 4.4)

$$H = (\text{part of } H \text{ inside } M) \cup (\text{part of } H \text{ inside } F)$$

i.e.,

$$H = (H \cap M) \cup (H \cap F) \tag{4.4}$$

(one of the two intersections can be empty). Recall that we have seen this before —in Example 4.4, the event H represented the incidence of a heart attack.

FIGURE 4.4

Partition of H into $H \cap M$ and $H \cap F$

Because $H \cap F$ and $H \cap M$ are mutually exclusive, (4.4) implies that

$$P(H) = P(H \cap F) + P(H \cap M)$$

Formulas (4.2) and (4.3) allow us to rewrite the two terms on the right side using conditional probabilities:

$$P(H) = P(H \mid F)P(F) + P(H \mid M)P(M) \qquad (4.5)$$

In this way, we have expressed the probability of H as a combination of the probabilities of the two sets M and F that form the partition of S. An easy way to remember this is to form a tree diagram (see Figure 4.5). Start with the sample space S and branch it out into the events that form the partition (M and F) and assign probabilities to each branch. Create further branches, and assign (this time, conditional) probabilities to them.

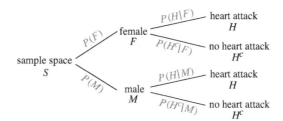

FIGURE 4.5

Tree diagram for $P(H)$

To find $P(H)$, we multiply the probabilities along each path leading to H and add up these products. In the same way we can calculate $P(H^c)$:

$$P(H^c) = P(H^c \mid F)P(F) + P(H^c \mid M)P(M) \qquad (4.6)$$

Formulas (4.5) and (4.6) are special cases of an important law in probability theory that we now discuss.

Definition 13 Partition

Let S be a sample space and $E_1, E_2, \ldots, E_n$ ($n \geq 1$) be events in S. If

(i) $E_1, E_2, \ldots, E_n$ are mutually exclusive, i.e.,

$$E_i \cap E_j = \emptyset \quad \text{for all} \quad i \neq j, \ 1 \leq i, j \leq n$$

(ii) $E_1, E_2, \ldots, E_n$ are collectively exhaustive, i.e.,

$$S = E_1 \cup E_2 \cup \cdots \cup E_n$$

then we say that $E_1, E_2, \ldots, E_n$ form a *partition* of the sample space S.

In many situations (data collection for research, or conducting surveys), we need to partition the sample set.

Example 4.8 Partitions

To study medical conditions related to smoking among adult Canadians (that's the sample space S), we partition S into smokers and non-smokers (the two sets are mutually exclusive and collectively exhaustive). Or, perhaps, we might need to partition into female smokers, male smokers, female non-smokers, and male non-smokers.

Note that we have already used the partition of a population into males and females in Example 4.4.

In researching drinking habits within a certain population, we might need to form a partition consisting of non-drinkers, occasional drinkers, light drinkers, moderate drinkers, and heavy drinkers.

By breaking down the set of genotypes $S = \{\text{RR}, \text{RG}, \text{GG}\}$ into homozygous and heterozygous, we formed a partition of S.

Example 4.9   **What Is Not a Partition?**

Consider again the sample space of genotypes $S = \{RR, RG, GG\}$. The events A = "kitten has at least one R allele" and B = "kitten has at least one G allele" are not mutually exclusive, since $A \cap B = \{RG\}$. Thus, A and B do not form a partition of S. Note, however, that A and B are collectively exhaustive, since a kitten must have at least one R allele or at least one G allele.

The sample space of the experiment that consists of tossing a coin three times in a row is

$$S = \{HHH, HHT, HTH, THH, HTT, THT, TTH, TTT\}$$

Define the events A = "outcome of the first toss is H," B = "outcome of the second toss is H," C = "outcome of the third toss is H," and D = "at least one T."

Note that $S = A \cup B \cup C \cup D$ but the sets are not mutually exclusive. For instance, $A \cap B = \{HHH, HHT\}$ or $B \cap D = \{THT, THH, HHT\}$. Thus, the four events do not form a partition of S.   ◣

Theorem 5   Law of Total Probability

Assume that the events $E_1, E_2, \ldots, E_n$ ($n \geq 1$) form a partition of a sample space S. For any event A in S,

$$P(A) = P(A \mid E_1)P(E_1) + P(A \mid E_2)P(E_2) + \cdots + P(A \mid E_n)P(E_n)$$ ◣

We have already shown in (4.5) that the theorem holds if the partition consists of two sets. A proof for a partition consisting of any number of sets is discussed in Exercise 31. A tree diagram for Theorem 5 is analogous to the one in Figure 4.5 (see also Example 4.10 and Figure 4.6).

Example 4.10   **Incidence of Rabies**

We study the incidence of rabies in an ecosystem. The population S of animals that can be infected with rabies consists of rabbits, foxes, and wolves. Table 4.1 contains the information about the distribution of animals within the population and the chance that an animal becomes infected with rabies.

Table 4.1

Animals	Percent of population	Probability of infection
rabbits	65	0.05
foxes	25	0.35
wolves	10	0.2

What is the probability that rabies will appear within the population S?

▶ Define the event R = "rabies present." We partition the sample space S into three sets—rabbits, foxes, and wolves—and create a tree diagram; see Figure 4.6.

By Theorem 5, the probability that rabies will appear in the population is (look at the tree diagram)

$$P(R) = (0.65)(0.05) + (0.25)(0.35) + (0.1)(0.2) = 0.14$$

We can do this more formally, to see exactly how Theorem 5 works.

Define the events E_1 = "rabbit," E_2 = "fox," and E_3 = "wolf." Clearly, E_1, E_2, and E_3 are mutually exclusive and collectively exhaustive, and therefore form a partition of S. It is given that $P(E_1) = 0.65$, $P(E_2) = 0.25$, and $P(E_3) = 0.1$. As well, from Table 4.1 we read $P(R \mid E_1) = 0.05$, $P(R \mid E_2) = 0.35$, and $P(R \mid E_3) = 0.2$.

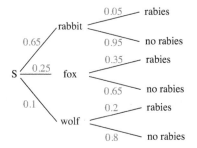

FIGURE 4.6

Tree diagram for the rabies infection

Using the law of total probability formula, we get

$$P(R) = P(\text{rabies} \mid \text{rabbit})P(\text{rabbit}) + P(\text{rabies} \mid \text{fox})P(\text{fox})$$
$$+ P(\text{rabies} \mid \text{wolf})P(\text{wolf})$$
$$= P(R \mid E_1)P(E_1) + P(R \mid E_2)P(E_2) + P(R \mid E_3)P(E_3)$$
$$= (0.05)(0.65) + (0.35)(0.25) + (0.2)(0.1) = 0.14$$

Example 4.11   **Kittens, Once Again**

Consider a sample of red-eyed cats, 30% of which are of genotype RG, and the remaining 70% of genotype RR. A cat from this group and a cat of genotype GG have a kitten. What is the probability that the kitten will have green eyes?

▶ Partition the sample space into two events: $E_1 = $ "RG and GG parents" and $E_2 = $ "RR and GG parents." Define $A = $ "kitten has green eyes."

It is given that $P(E_1) = 0.3$ and $P(E_2) = 0.7$. In order to use the law of total probability

$$P(A) = P(A \mid E_1)P(E_1) + P(A \mid E_2)P(E_2)$$

we need to figure out the two conditional probabilities.

The two possible genotypes of a kitten from RG and GG parents are RG and GG, each occurring with a 50% chance. The RG trait is red eyes, whereas GG yields green eyes. We conclude that

$$P(\text{green eyes} \mid \text{RG and GG parents}) = P(A \mid E_1) = 0.5$$

The only genotype of a kitten whose parents are RR and GG is RG, which yields red eyes. Consequently,

$$P(\text{green eyes} \mid \text{RR and GG parents}) = P(A \mid E_2) = 0$$

We are done—the probability that the kitten has green eyes is

$$P(A) = P(A \mid E_1)P(E_1) + P(A \mid E_2)P(E_2)$$
$$= (0.5)(0.3) + (0)(0.7) = 0.15$$

What is the probability of the event $B = $ "kitten has red eyes"? Since B is complementary to A, we get $P(B) = 1 - P(A) = 1 - 0.15 = 0.85$.

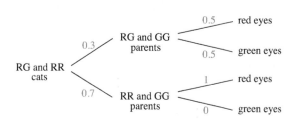

FIGURE 4.7

Tree diagram for $P(B)$

To practise total probabilities, we construct a tree diagram (see Figure 4.7) and figure out $P(B)$ from there.

Following the two paths that lead to red eyes, we get

$$P(B) = (0.5)(0.3) + (1)(0.7) = 0.85$$

Bayes' Theorem

By combining the properties we have seen so far of the probability function, we discover a useful formula.

Assume that the events E_1 and E_2 form a partition of a sample space S. Let A be an event and assume that the conditional probabilities $P(A \mid E_1)$ and $P(A \mid E_2)$ are known. How do we compute $P(E_1 \mid A)$ and $P(E_2 \mid A)$?

By the definition of conditional probability, we have

$$P(E_1 \mid A) = \frac{P(E_1 \cap A)}{P(A)} \tag{4.7}$$

Since $P(A \mid E_1)$ is known, we use (4.2):

$$P(E_1 \cap A) = P(A \cap E_1) = P(A \mid E_1)P(E_1) \tag{4.8}$$

For the denominator of (4.7) we use Theorem 5:

$$P(A) = P(A \mid E_1)P(E_1) + P(A \mid E_2)P(E_2) \tag{4.9}$$

Substituting (4.8) and (4.9) into (4.7), we get

$$P(E_1 \mid A) = \frac{P(E_1 \cap A)}{P(A)} = \frac{P(A \mid E_1)P(E_1)}{P(A \mid E_1)P(E_1) + P(A \mid E_2)P(E_2)}$$

In exactly the same way, we obtain a formula for $P(E_2 \mid A)$.

In the general case of a partition of S into n events, we replace the denominator by the formula from Theorem 5. Thus, we have proved the following theorem.

Theorem 6   Bayes' Formula

Assume that the events $E_1, E_2, \ldots, E_n$ ($n \geq 1$) form a partition of a sample space S. Let A be an event. Then

$$P(E_i \mid A) = \frac{P(A \mid E_i)P(E_i)}{P(A \mid E_1)P(E_1) + P(A \mid E_2)P(E_2) + \cdots + P(A \mid E_n)P(E_n)}$$

for $i = 1, 2, \ldots, n$.

Why is this useful? Consider the following examples.

Example 4.12   **Kittens, for the Last Time (in This Section)**

We continue with Example 4.11. A randomly chosen kitten has green eyes. What is the probability that its parents are RG and GG?

▶ We are asked to calculate $P(E_1 \mid A)$. We use Bayes' formula,

$$P(E_1 \mid A) = \frac{P(A \mid E_1)P(E_1)}{P(A \mid E_1)P(E_1) + P(A \mid E_2)P(E_2)}$$

$$= \frac{(0.5)(0.3)}{(0.5)(0.3) + (0)(0.7)} = 1$$

Hardly a surprise: RR and GG parents can only have an RG kitten, which means that they cannot have a kitten with green eyes. So from $P(E_2 \mid A) = 0$ we conclude that $P(E_1 \mid A) = 1$.

Example 4.13 **Testing for Breast Cancer**

Among the sample space of women aged $40-50$, the prevalence of breast cancer is 0.8%. A test for the presence of breast cancer (say, a mammogram) shows a positive result in 90% of women who have breast cancer and in 5% of women who do not have breast cancer (these are called false positives). We will answer the following questions:

(1) What is the probability that the test result is positive for a randomly chosen woman from the sample space?

(2) Here is a more important (more relevant) question—suppose that a woman takes the test and the test shows a positive result. How likely is it that she has breast cancer?

▶ As usual, we define events first and then translate the given information into probabilities.

(1) We partition the population S into two sets: $E_1 =$ "a woman has breast cancer" and $E_2 =$ "a woman does not have breast cancer." Let $A =$ "test shows a positive result."

It is given that $P(E_1) = 0.008$. Thus, $P(E_2) = 1 - P(E_1) = 0.992$. Within the group of women who have breast cancer, the test turns out positive in 90% of the cases: $P(A \mid E_1) = 0.9$. Therefore, $P(A^C \mid E_1) =$ the probability that a woman who has breast cancer tests negative (this situation is called a false negative) is 0.1.

The false-positive information means that $P(A \mid E_2) = 0.05$. The probability $P(A^C \mid E_2) = 0.95$ represents true negatives.

To visualize all of this, we construct a tree diagram; see Figure 4.8.

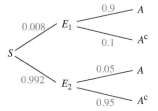

FIGURE 4.8

Tree diagram for $P(A)$

Looking at the tree, we see that two paths lead to A; thus,
$$P(A) = (0.008)(0.9) + (0.992)(0.05) = 0.0568$$
Formally, using the law of total probability
$$P(A) = P(A \mid E_1)P(E_1) + P(A \mid E_2)P(E_2)$$
$$= (0.9)(0.008) + (0.05)(0.992) = 0.0568$$

Thus, the probability that a randomly chosen woman from the given sample tests positive for brest cancer is 5.68%.

(2) We are asked to calculate $P(E_1 \mid A)$, i.e., the probability that a woman who tests positive for breast cancer (that's the condition) actually has breast cancer.

We know the event A, and the probability we are looking for is one of the partition events conditional on A. This is the situation that requires Bayes' formula. By Theorem 6,
$$P(E_1 \mid A) = \frac{P(A \mid E_1)P(E_1)}{P(A \mid E_1)P(E_1) + P(A \mid E_2)P(E_2)}$$
$$= \frac{(0.9)(0.008)}{(0.9)(0.008) + (0.05)(0.992)} \approx 0.1268 \qquad (4.10)$$

So the probability is a bit larger than 12.5%.

The reason the probability is so small is that the chance that a woman has breast cancer is small in the first place (0.8%). In practice, if a single test gives a positive result, additional testing is done to confirm or dismiss the result of the test.

Let's think a bit more about this. If the prevalence of breast cancer in the general population were 8% (instead of 0.8%), then (replace 0.008 by 0.08 and 0.992 by 0.92 in (4.10))

$$P(E_1 \mid A) = \frac{P(A \mid E_1)P(E_1)}{P(A \mid E_1)P(E_1) + P(A \mid E_2)P(E_2)}$$
$$= \frac{(0.9)(0.08)}{(0.9)(0.08) + (0.05)(0.92)} \approx 0.6102$$

Thus, the probability that a woman who tested positive for breast cancer actually has breast cancer has increased dramatically.

To see this better, assume that the prevalence of breast cancer in the general population is p, where $0 \le p \le 1$. The tree diagram for the probabilities is shown in Figure 4.9.

In this case,

$$P(E_1 \mid A) = \frac{P(A \mid E_1)P(E_1)}{P(A \mid E_1)P(E_1) + P(A \mid E_2)P(E_2)}$$
$$= \frac{0.9p}{0.9p + 0.05(1 - p)} = \frac{0.9p}{0.85p + 0.05}$$

When $p = 0.008$, then $P(E_1 \mid A) = 0.1268$, and when $p = 0.08$, then $P(E_1 \mid A) = 0.6102$, confirming our earlier calculations. The graph of $P(E_1 \mid A)$ as a function of p (see Figure 4.10) shows the initial steep increase in $P(E_1 \mid A)$.

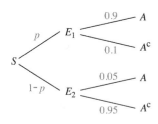

FIGURE 4.9

Tree diagram to help us compute $P(E_1 \mid A)$

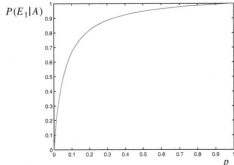

FIGURE 4.10

Dependence of $P(E_1 \mid A)$ on p

To check our answer to (2), we can argue using relative frequencies. Take a sample of 10,000 women aged 40–50. Since the prevalence of breast cancer in this group is 0.8%, we conclude that 80 women have breast cancer and 9,920 do not have breast cancer.

How many women will test positive for breast cancer? 5% of the 9,920 women (496 women) who do not have breast cancer will nevertheless test positive (these are false positives). Of the 80 women who have breast cancer, 90% (i.e., 72) will test positive.

So, a total of $496 + 72 = 568$ women will test positive. Of those, 72 have breast cancer. Therefore, the probability that a woman who tests positive has breast cancer is $72/568 \approx 0.1268$.

Summary   **Conditional probability** allows us to calculate the probability of an event occurring when we know that another event has taken place. Calculating conditional probability is based on basic properties of probability and set theory. A **partition** of a set is a collection of pairwise disjoint subsets whose union is the whole set (think about tiling a floor: no overlaps and no gaps). The **law of total probability** tells us how to calculate the probability of an event based on knowing the probabilities related to the sets that form a partition of the sample set. In this context, the conditional probabilities are calculated using **Bayes' formula.**

4	Exercises

1. Find a pair of events A and C in the sample space $S = \{1, 2, 3, 4, 5\}$ for which $P(A \mid C) = 1$. Find a pair of events B and D for which $P(B \mid D) = 0$.

2. Find a pair of events A and C in the sample space $S = \{1, 2, 3, 4, 5\}$ for which $P(A \mid C) = P(A)$. Find a pair of events B and D for which $P(B \mid D) = P(D)$.

▽ 3–6 ▪ The sample space is $S = \{1, 2, 3, 4, 5\}$. Find $P(A \cap B)$, $P(A \mid B)$, and $P(B \mid A)$.

3. $P(1) = 0.2$, $P(2) = 0.1$, $P(3) = 0.15$, $P(4) = 0.45$, $P(5) = 0.1$; $A = \{1, 2, 3\}$, $B = \{1, 4, 5\}$.

4. $P(1) = 0.2$, $P(2) = 0.1$, $P(3) = 0.15$, $P(4) = 0.45$, $P(5) = 0.1$; $A = \{1, 2\}$, $B = \{3, 4, 5\}$.

5. $P(1) = 0.1$, $P(2) = 0.3$, $P(3) = 0.2$, $P(4) = 0.3$, $P(5) = 0.1$; $A = \{1, 2, 4, 5\}$, $B = \{4, 5\}$.

◣ 6. $P(1) = 0$, $P(2) = 0.2$, $P(3) = 0.3$, $P(4) = 0.3$, $P(5) = 0.2$; $A = \{1, 2, 4\}$, $B = \{3, 4, 5\}$.

7. Let $S = \{1, 2, 3, 4, 5\}$ be a sample space, and assume that all five events are equally likely to occur.

   (a) Find two sets A and B so that $P(A \mid B) \neq P(B \mid A)$.

   (b) Find two sets A and B so that $P(A \mid B) = P(B \mid A)$.

8. Let $S = \{1, 2, 3, 4, 5\}$ be a sample space, and assume that $P(1) = 0.15$, $P(2) = 0.25$, $P(3) = 0.2$, $P(4) = 0.3$, and $P(5) = 0.1$.

   (a) Find two sets A and B so that $P(A \mid B) \neq P(B \mid A)$.

   (b) Find two sets A and B so that $P(A \mid B) = P(B \mid A)$.

▽ 9–15 ▪ Using *conditional probability,* answer the following questions. If you wish, check your answers by using other means.

9. A family has three children, two of which are girls. Assuming a 1:1 sex ratio of births, what is the probability that their third child is a boy?

10. A family has four children. Knowing that three of them are boys and assuming a 1:1 sex ratio of births, what is the probability that the fourth child is a girl?

11. A family has four children. Knowing that three are of the same sex and assuming a 1:1 sex ratio of births, what is the probability that the fourth child is a girl?

12. A coin is tossed three times. Find the probability that exactly two heads occurred given that at least one toss resulted in heads.

13. A coin is tossed three times. Find the probability that at least two heads occurred given that at least one toss resulted in heads.

14. Two dice are rolled. Find the probability that one die is a 4 given that the sum is 6.

15. Two dice are rolled. Find the probability that the sum is 7 given that one die shows a number larger than 3.

▽ 16–18 ▪ Within a population of tigers, the trait of the dominant allele P is a spotted tail, and the trait of the recessive allele T is a striped tail.

16. A baby tiger, born to genotype PT parents, has a spotted tail. What is the probability that it is of genotype PT?

17. A baby tiger, born to genotype PT parents, is known to have one T allele. What is the probability that it has a striped tail?

18. A baby tiger, born to genotype PT and TT parents, is known to have one T allele. What is the probability that it has a striped tail?

19. Let $S = \{1, 2, 3, 4, 5\}$.

    (a) Find three subsets of S whose union is S but which do not form a partition of S.

    (b) Find three mutually exclusive subsets of S that do not form a partition of S.

    (c) Find a partition of S that contains three sets.

20. Let $S = \{1, 2, 3, 4\}$.

    (a) Find three mutually exclusive subsets of S that do not form a partition of S.

    (b) Find a collection of two subsets of S whose union is S but that are not mutually exclusive.

    (c) Find a partition of S that contains three sets.

21. A surveyed population consists of 60% females and 40% males. Of all males, 35% are smokers, and of all females, 20% are smokers.

    (a) What is the probability that a randomly chosen person from this population is a smoker?

    (b) What is the probability that a randomly selected smoker is male?

22. Within a population with a 1:1 sex ratio, 15% of females have blond hair and 5% of males have blond hair.

    (a) What is the probability that a randomly chosen person has blond hair?

    (b) What is the probability that a randomly selected blond person is female?

23. A certain population consists of 20% children, 30% adolescents, and 50% adults. The probabilities that a member of this population catches the flu are 0.45 for a child, 0.2 for an adolescent, and 0.15 for an adult.

    (a) What is the probability that a randomly selected member of this population has the flu?

    (b) What is the probability that a randomly selected person with the flu is an adult?

24. Of all frogs in a large pond, 50% are green, 35% are brown, and 15% are blue. About 5% of green frogs have brown eyes, and 75% of blue frogs have brown eyes. There are no brown frogs with brown eyes.

    (a) What is the probability that a randomly chosen frog has brown eyes?

    (b) A randomly selected frog has been found to have brown eyes. What is the probability that it is green?

25. The incidence of asthma in young adults (assuming a 1:1 sex ratio) is 6.4% for females and 4.5% for males. [Source: Thomsen, S.F., Ulrik, C.S., Kyvik, K.O., Larsen, K., Skadhauge, L.R., Steffensen, I., et al. (2005). The incidence of asthma in young adults. *Chest*, 127 (6), 1928-1934.]

    (a) What is the probability that a randomly chosen young adult has asthma?

(b) What is the probability that a young adult with asthma is a female?

26. A certain medical condition (could be high blood pressure) comes in three forms, X, Y, and Z, with prevalences of 45%, 35%, and 20%, respectively. The probability that a person will need emergency medical attention is 10% if he has the X form, 5% if he has the Y form, and 45% if he has the Z form. What is the probability that a person who has the condition will require emergency medical attention?

27. There is an open-air concert tomorrow, but the weather forecast does not look good: a 60% chance of rain. You need transportation to go to the concert and know the following: if it rains, there is a 30% chance that you will have a car available, but if it does not rain, the chance is 90%. How likely is it that you will go to the concert tomorrow?

28. There is a way to survey people face-to-face on embarrassing questions and ensure their anonymity (thus getting potentially useful results). Suppose you live in Toronto and wish to survey people on their use of a cell phone while driving (which is against the law in Ontario). You set it up in the following way. Each participant draws a card from a deck of cards; you don't know which card they picked. If they pick a black suit, then they have to truthfully answer a question where there is exactly a 50-50 chance of answering "yes" or "no" (such as "Was your mother born between 1 January and 30 June?"). If they pick a red suit, then they have to truthfully answer the question "Have you driven a car and talked on a cell phone at the same time?"

   You realize that of the 200 people you surveyed, 120 answered "yes." So, how many people are likely to use their cell phone while driving?

29. The incidence of bacterial meningitis in Canada ranged from 3.17 to 3.66 per 100,000 between 1994 and 2001. [Source: Public Health Agency of Canada, http://www.phac-aspc.gc.ca/publicat/ccdr-rmtc/05vol31/dr3123a-eng.php.] For the purpose of this exercise, we take the incidence to be 3.4 per 100,000. A test for meningitis shows a positive result in 85% of people who have it and in 7% of people who do not have it.

   (a) What is the probability that a randomly selected person tests positive for bacterial meningitis?

   (b) If a person tests positive for bacterial meningitis, what is the probability that they have it?

30. The average incidence of autism spectrum disorder is 45 cases per $10,000$. [Source: Rutter, M. (2005). Incidence of autism spectrum disorders: changes over time and their meaning. *Acta Paediatrica*, 94 (1), 2-15.] A test for the disorder shows a positive result in 96% of people who have the disorder, and in 1% of people who do not have it.

   (a) What is the probability that a randomly selected person tests positive for the disorder?

   (b) If a person tests positive for the disorder, what is the probability that they have it?

31. Assume that the sets E_1, E_2, and E_3 form a partition of a sample space S and let A be an event in S.

   (a) Using Venn diagrams, show that A can be written as a union of $A \cap E_1$, $A \cap E_2$, and $A \cap E_3$. Explain why the formula $P(A) = P(A \cap E_1) + P(A \cap E_2) + P(A \cap E_3)$ is true.

   (b) Show that (a) implies that $P(A) = P(A \mid E_1)P(E_1) + P(A \mid E_2)P(E_2) + P(A \mid E_3)P(E_3)$, which is the law of total probability in the case $n = 3$.

   (c) Repeat (a) and (b) in the case of a partition of S into n subsets.

32. Recall that the events $E_1, E_2, \ldots, E_n$ are called mutually exclusive if the condition $E_i \cap E_j = \emptyset$ holds for all i, j between 1 and n. How many conditions do we need to check if $n = 3$, $n = 4$, and $n = 5$? In general, for any n?

5   Independence

We introduce an important category of events, called **independent events.** If our knowledge about an event does not tell us anything about the probability of another event occurring, then the two events are independent.

Independent Events

We toss a coin and it's tails. We toss it again. Knowing that it was tails the first time, what is the probability that it will be tails again?

Define the events A = "outcome of the second toss is T" and B = "outcome of the first toss is T." The question is, what is $P(A \mid B)$? Since the coin has no memory, the outcome of the second toss does not depend on the outcome of the first toss. Thus,

$$P(A \mid B) = P(A) \tag{5.1}$$

If (5.1) holds for two events A and B, then we say that A is *independent* of B. Combining the definition of conditional probability

$$P(A \mid B) = \frac{P(A \cap B)}{P(B)}$$

with (5.1), we obtain

$$\frac{P(A \cap B)}{P(B)} = P(A)$$

and

$$P(A \cap B) = P(A)P(B) \tag{5.2}$$

Definition 14   Independent Events

Two events A and B are called *independent* if $P(A \cap B) = P(A)P(B)$.

Although (5.1) and Definition 14 are equivalent, the statement given in Definition 14 is a common way of defining independence. The major reason is that the definition emphasizes *symmetry:* note that switching A and B in $P(A \cap B) = P(A)P(B)$ results in the same formula. Thus, we do not have to say "A is independent of B" (as in $P(A \mid B) = P(A)$) nor "B is independent of A" (as in $P(B \mid A) = P(B)$); instead, we say that "A and B are independent."

Example 5.1   Independent Events: Rolling a Die

What is the probability that two consecutive rolls of a die result in two sixes?

▶ Define the events A_1 = "outcome of the first roll is a 6" and A_2 = "outcome of the second roll is a 6." We know that $P(A_1) = P(A_2) = 1/6$. Since A_1 and A_2 are independent,

$$P(A_1 \cap A_2) = P(A_1)P(A_2) = \frac{1}{6} \cdot \frac{1}{6} = \frac{1}{36}$$

Note that $A_1 \cap A_2$ = "outcome of both rolls is a 6," and so the answer to the question is 1/36.

Example 5.2   Independence: Genetics

Both the mother and the father are of genotype RG. What is the probability that their offspring is of genotype GG?

▶ One way to do this is to define the sample space of all possible genotypes $S = \{RR, RG, GR, GG\}$ and look at probabilities. Since the four events are equally likely, $P(GG) = 1/4$.

Here is an alternative way of thinking about this: in order for an offspring to inherit the GG genotype, it has to inherit the allele G from its mother and the allele G from its father. Define the events $A_1 =$ "allele from the mother is G," $A_2 =$ "allele from the father is G," and $B = A_1 \cap A_2 =$ "offspring has genotype GG."

From biological laws (laws of inheritance) we know that $P(A_1) = P(A_2) = 1/2$ and that the events A_1 and A_2 are independent. Thus,

$$P(B) = P(A_1 \cap A_2) = P(A_1)P(A_2) = \frac{1}{2} \cdot \frac{1}{2} = \frac{1}{4}$$ ◢

The definition of independence can be extended to any number of events. Although the message is the same, the actual conditions get messy. In particular, three events A, B, and C are said to be independent if *all* of the following conditions hold: $P(A \cap B) = P(A)P(B)$, $P(A \cap C) = P(A)P(C)$, $P(B \cap C) = P(B)P(C)$, and $P(A \cap B \cap C) = P(A)P(B)P(C)$.

To define independence for n events, we have to check all possible combinations of intersections of $2, 3, 4, \ldots, n$ events (see Exercise 23). Thus, deciding whether or not several events are independent using the definition could be quite hard. However, if we know that events are independent, then we can calculate the probability of their intersection.

Example 5.3   Using Independence

Assume that a certain population has a 1:1 sex ratio of births.

(a) A family has three children. What is the probability that they are all girls?

(b) A family has three children. What is the probability that at least one child is a boy?

▶ (a) Define the following events: $A_1 =$ "the first child is a girl," $A_2 =$ "the second child is a girl," and $A_3 =$ "the third child is a girl." Girls and boys are equally likely to be born, so $P(A_1) = P(A_2) = P(A_3) = 1/2$. Define $B = A_1 \cap A_2 \cap A_3 =$ "all three children are girls." The fact that the consecutive births are considered independent implies that

$$P(B) = P(A_1 \cap A_2 \cap A_3) = P(A_1)P(A_2)P(A_3) = \left(\frac{1}{2}\right)^3 = \frac{1}{8}$$

(b) Here is one approach: the sample space of all events is

$$S = \{GGG, GGB, GBG, GBB, BBB, BBG, BGB, BGG\}$$

The sequences of three letters (B for a boy, G for a girl) represent the chronological order of births. The probabilities are $P(\text{three girls}) = P(\text{three boys}) = 1/8$, and $P(\text{two boys and a girl}) = P(\text{two girls and a boy}) = 3/8$. It follows that

$$P(\text{at least one boy}) = P(\text{exactly one boy}) + P(\text{exactly two boys}) + P(\text{three boys})$$

$$= \frac{3}{8} + \frac{3}{8} + \frac{1}{8} = \frac{7}{8}$$

A commonly used trick — to consider the complementary event — simplifies our calculations quite a bit: instead of the event "at least one boy," consider the complement "no boys" = "all girls." From (1) we know that $P(\text{all girls}) = 1/8$, and therefore $P(\text{at least one boy}) = 1 - P(\text{all girls}) = 7/8$.

Example 5.4   Independence in Genetics: Example of the Hardy-Weinberg Law

Assume that, within some population, the fraction of G alleles is 0.6 and the fraction of R alleles is 0.4 (this means that the fraction of G alleles among females is 0.6 and the fraction of G alleles among males is 0.6; the fraction of R alleles among the females, as well as among the males, is 0.4).

To determine the genetic makeup of a child, we select one allele from the mother and one allele from the father. These two selections are considered independent. To guide us through the reasoning, we construct a tree diagram; see Figure 5.1.

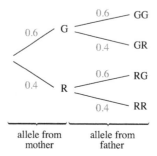

FIGURE 5.1

Tree diagram for allele selection

Using the assumption on independence,

$$P(\text{GG}) = P(\text{G from mother and G from father})$$
$$= P(\text{G from mother})P(\text{G from father}) = 0.6 \cdot 0.6 = 0.36$$

Likewise,

$$P(\text{GR}) = P(\text{G from mother and R from father})$$
$$= P(\text{G from mother})P(\text{R from father}) = 0.6 \cdot 0.4 = 0.24$$

In the same way, we calculate $P(\text{RG}) = 0.24$ and $P(\text{RR}) = 0.16$.

Suppose that 1,000 children are born. What is the genetic makeup of the new generation?

We use the probabilities we just found: of the 1,000 children, $0.36(1,000) = 360$ will be of type GG, $0.24(1,000) + 0.24(1,000) = 480$ will be of type GR = RG, and $0.16(1,000) = 160$ will be of type RR.

The 360 genotype GG children contribute 720 G alleles to the genetic pool. The 480 genotype RG children contribute 480 G and 480 R alleles to the pool, and the 160 genotype RR children contribute 320 R alleles to the pool. The genetic pool of children will contain 2,000 alleles, of which $720 + 480 = 1,200$ are G alleles and $480 + 320 = 800$ are R alleles. The ratio of G alleles is $1,200/2,000 = 0.6$, and the ratio of R alleles is $800/2,000 = 0.4$. Note that the ratio is the same as in the genetic pool of the parents.

This example is an illustration of the *Hardy-Weinberg law*, which states that in the absence of other factors, the genetic makeup of a population remains unchanged.

Example 5.5   Independence in Genetics: Hardy-Weinberg Law

Now we generalize the calculations of Example 5.4. Assume that the fraction of G alleles within some population is g and the fraction of R alleles is r (thus, $g + r = 1$). The tree diagram in Figure 5.2 shows all possibilities for the genetic makeup of a child, and the corresponding probabilities.

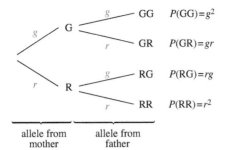

FIGURE 5.2

Tree diagram for the genetic makeup of a child

Suppose that the population has N children. What are the fractions of G and R alleles in this new genetic pool?

We organize the information in Table 5.1.

Table 5.1

Genotype	Number of children	Number of G alleles	Number of R alleles
GG	$g^2 N$	$2g^2 N$	0
GR, RG	$2grN$	$2grN$	$2grN$
RR	$r^2 N$	0	$2r^2 N$

Thus, there is a total of

$$2g^2 N + 2grN = 2gN(g+r) = 2gN$$

G alleles and

$$2grN + 2r^2 N = 2rN(g+r) = 2rN$$

R alleles (keep in mind that $g + r = 1$). The total number of alleles is $2N$; the ratio of G alleles is $2gN/2N = g$ and the ratio of R alleles is $2rN/2N = r$. Thus, the ratios have not changed.

As an illustration of situations that produce non-independent events we consider an example of a *Markov chain*. Recall that (see Example 2.2 in Section 2 and the text following it) a Markov chain is a stochastic system where the probability of arriving at a particular state depends on the state at the previous time.

Example 5.6   Markov Chain: Disappearance and Recurrence of a Virus

Recall the model of recurrence and disappearance of a virus from Example 2.2: if the virus is present in the population at time t, it will be present at time $t+1$ with a probability of 0.75. If it is absent from the population at time t, the virus will appear in the population at time $t+1$ with a probability of 0.2; see Figure 5.3.

FIGURE 5.3

Dynamics of disappearance and recurrence of the virus

Define the following events:

V_t = "virus present at time t"

N_t = "virus absent at time t"

V_{t+1} = "virus present at time $t+1$"

N_{t+1} = "virus absent at time $t+1$"

The assumptions about the behaviour of the virus can be translated into conditional probabilities in the following way:

$$P(V_{t+1} \mid V_t) = 0.75$$
$$P(V_{t+1} \mid N_t) = 0.2$$
$$P(N_{t+1} \mid N_t) = 0.8$$
$$P(N_{t+1} \mid V_t) = 0.25$$

We now compute the probability that the virus is present at time $t + 1$.

The events V_t and N_t form a partition of the sample space (the virus is either present or not). By the law of total probability,

$$P(V_{t+1}) = P(V_{t+1} \mid V_t)P(V_t) + P(V_{t+1} \mid N_t)P(N_t)$$
$$= 0.75P(V_t) + 0.2P(N_t)$$

Thus, the probability at time $t + 1$ depends on the probabilities at time t (which we do not know). From

$$P(V_{t+1} \mid V_t) = 0.75 \neq P(V_{t+1} \mid N_t) = 0.2$$

we conclude that V_{t+1} and V_t are not independent (it is more likely that V_{t+1} occurs if V_t occurs than if N_t occurs; see Exercise 24 for a formal proof).

To visualize the long-term behaviour, we run the simulation for 100 time steps; see Figure 5.4, where 0 marks the absence and 1 marks the presence of the virus. (Note that, due to the chance factor, the picture differs from the one in Figure 2.4 in Section 2.)

FIGURE 5.4

Dynamics of recurrence and disappearance of a virus, 100 steps

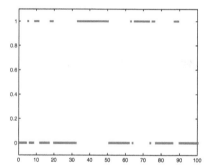

We contrast the situation of Example 5.6 with the situation where the probabilities of switching between the two states, A and B, are higher (see Figure 5.5).

FIGURE 5.5

The dynamics of high probability of switching between states

As before, we define the events

    A_t = "system is at A at time t"

    B_t = "system is at B at time t"

    A_{t+1} = "system is at A at time $t + 1$"

    B_{t+1} = "system is at B at time $t + 1$"

From the diagram, we see that

$$P(A_{t+1} \mid A_t) = 0.25$$
$$P(A_{t+1} \mid B_t) = 0.9$$

$$P(B_{t+1} \mid A_t) = 0.75$$
$$P(B_{t+1} \mid B_t) = 0.1$$

Again, the fact that $P(A_{t+1} \mid A_t) \neq P(A_{t+1} \mid B_t)$ shows that A_{t+1} and A_t are not independent.

To see the long-term behaviour, we run the simulation for 100 steps; see Figure 5.6 (0 represents state A and 1 represents state B).

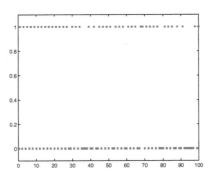

FIGURE 5.6

The dynamics of a high probability of switching between states

So, using simple stochastic tools, we can model various patterns of switching behaviours.

Example 5.7   **Mutation**

Assume that r_t and g_t represent the ratios of alleles R and G, respectively, in a reproductive population at time t. The Hardy-Weinberg law states that the genetic pool in the next generation, without any outside factors, will keep the same ratios.

Now consider the case of *mutation:* in each generation, the fraction m of G-alleles mutates to R-alleles.

Suppose that there are N offspring (thus, $2N$ alleles). Of the $2N$ alleles, $2Ng_t$ are G-alleles and $2Nr_t$ are R-alleles. Due to the mutation, $m(2Ng_t)$ G-alleles become R-alleles. Thus, of the $2N$ alleles in generation $t+1$, there are $2Nr_t + 2mNg_t$ R-alleles and $2Ng_t - 2mNg_t$ G-alleles.

The ratios are

$$r_{t+1} = \frac{2Nr_t + 2mNg_t}{2N} = r_t + mg_t$$

and

$$g_{t+1} = \frac{2Ng_t - 2mNg_t}{2N} = g_t - mg_t = (1 - m)g_t$$

From $r_t + g_t = 1$ we get $g_t = 1 - r_t$ and

$$r_{t+1} = r_t + m(1 - r_t) = (1 - m)r_t + m$$

Thus, we obtain the *deterministic* dynamical system for the ratios:

$$g_{t+1} = (1 - m)g_t$$
$$r_{t+1} = (1 - m)r_t + m$$

The solutions (see Exercise 25) are given by

$$g_t = g_0(1 - m)^t$$
$$r_t = (r_0 - 1)(1 - m)^t + 1 \tag{5.3}$$

where g_0 and r_0 are the ratios of the G and R alleles, respectively, in the starting generation.

Since $1 - m < 1$, the ratio of G-alleles decreases exponentially and

$$\lim_{t \to \infty} g_t = 0$$

On the other hand,

$$\lim_{t \to \infty} r_t = (r_0 - 1)(0) + 1 = 1$$

Thus, in the long term, G-alleles will completely disappear from the population.
Substituting $g_t = 0.5g_0$ into (5.3), we calculate the half-life

$$0.5 = (1 - m)^t$$
$$\ln 0.5 = t \ln(1 - m)$$
$$t = \frac{\ln 0.5}{\ln(1 - m)}$$

Mutation rates are usually very small. If $m = 0.001$, then the half-life of G-alleles is

$$t = \frac{\ln 0.5}{\ln(1 - 0.001)} \approx 693$$

generations.

Summary   Two events are **independent** if the knowledge of one event does not convey anything about the probability of the other event. In other words, if the probabilify of one event conditional on another is equal to the unconditional probability, then the two events are independent. For independent events, the probability of the intersection of events is equal to the product of the individual probabilities. The **Hardy-Weinberg law** states that, in the absence of outside factors, the genetic makeup of a population remains unchanged.

5   Exercises

1. Assume that A and B are disjoint events such that $P(A) > 0$ and $P(B) > 0$. Can A and B be independent?

2–5 ▪ Define the sample space $S = \{1, 2, 3, 4, 5\}$, and assume that $P(1) = 0.2$, $P(2) = 0.1$, $P(3) = 0.2$, $P(4) = 0.4$, and $P(5) = 0.1$.

2. Let $A = \{1, 3\}$ and $B = \{2, 4\}$. Calculate $P(A \mid B)$ and $P(A)$. Are A and B independent?

3. Let $A = \{4, 5\}$ and $B = \{2, 5\}$. Calculate $P(B \mid A)$ and $P(B)$. Are A and B independent?

4. Let $A = \{1\}$ and $B = \{1, 2, 3, 4, 5\}$. Calculate $P(B \mid A)$ and $P(B)$. Are A and B independent?

5. Let $A = \{1, 5\}$ and $B = \{1, 2, 5\}$. Calculate $P(A \mid B)$ and $P(A)$. Are A and B independent?

6. The sample space consists of four elements, $S = \{1, 2, 3, 4\}$. Assume that $P(1) = P(2) = P(3) = 0.2$, and $P(4) = 0.4$. Are the events $A = \{1, 4\}$ and $B = \{2, 3, 4\}$ independent?

7. The sample space consists of four elements, $S = \{1, 2, 3, 4\}$. Assume that $P(1) = 0.2$, $P(2) = 0.3$, $P(3) = 0.2$, and $P(4) = 0.3$. Are the events $A = \{1, 3\}$ and $B = \{1, 4\}$ independent?

8. The sample space consists of four elements, $S = \{1, 2, 3, 4\}$. Assume that $A = \{2, 3\}$ and $B = \{3, 4\}$ are independent and $P(A) = 0.3$ and $P(B) = 0.2$. Find $P(3)$ and $P(1)$.

9. The sample space consists of four elements, $S = \{1, 2, 3, 4\}$. Assume that $A = \{1, 3\}$ and $B = \{2, 3\}$ are independent and $P(A) = 0.5$ and $P(B) = 0.4$. Find $P(3)$ and $P(4)$.

10. The sample space consists of four elements, $S = \{1, 2, 3, 4\}$. Assume that $P(1) = 0.1$, $P(2) = 0.4$, $P(3) = 0.1$, and $P(4) = 0.4$, and let $A = \{1, 3\}$. Find an event B that contains two elements and is independent of A.

11. The sample space consists of four elements, $S = \{1, 2, 3, 4\}$. Assume that $P(1) = 0.3$, $P(2) = 0.2$, $P(3) = 0.2$, and $P(4) = 0.3$, and let $A = \{2, 4\}$. Find an event B that contains two elements and is independent of A.

12. A quiz has five multiple-choice questions, each with three choices. Without reading them, a student randomly answers all questions.

    (a) What is the probability that the student will answer at least one question correctly?

    (b) What is the probability that the student will answer all questions correctly?

13. A quiz has ten multiple-choice questions, each with two choices. Without reading them, a student randomly answers all questions.

    (a) What is the probability that the student will answer at least one question correctly?

    (b) What is the probability that the student will answer all questions correctly?

14. The probability that a couple has a female child is 0.54 and the probability that they have a male child is 0.46. Assume that births are independent events.

    (a) A couple has three children. What is the probability that exactly two of their children are girls?

    (b) A couple has four children. What is the probability that at least two children are boys?

15. The probability that a couple has a female child is 0.45 and the probability that they have a male child is 0.55. Assume that births are independent events.

    (a) A couple has three children. What is the probability that exactly two of their children are girls?

    (b) A couple has four children. What is the probability that at least two children are boys?

16. The chance that a mosquito survives an application of a bug spray is 0.15. What is the probability that in a sample of 20 mosquitoes at least one will survive?

17. One way to get rid of most of the house dust mites (which are the most commom cause of allergic reactions and asthma) is to wash laundry in hot water. It has been determined that the chance that a house dust mite survives in laundry washed at 60°C is 0.01. What is the probability that, in a sample of 100 house dust mites, at least one will survive?

18. A medical test for a certain disease gives a false-positive result with a probability of 0.002. (A false positive describes the situation where the test turns out positive although the person tested does not have the disease.) What is the probability that in a group of 100 people, at least one false positive will occur?

19. A medical test for high blood glucose gives a false-negative result with a probability of 0.012. (A false negative describes the situation where the test turns out negative although the person tested has a high blood glucose level.) What is the probability that in a group of 50 people, at least one false negative will occur?

20. The average efficacy of an oral contraceptive (birth control pill) is about 97.5% per year. This means that, within a year, 2.5% of sexually active women who are taking the pill will get pregnant. What is the probability that a sexually active woman who takes birth control pills will get pregnant at least once in a 5-year period?

21. The average efficacy of a condom (used without any other preventative measures) is about 86% per year. This means that, within a year, 14% of sexually active women who use condoms will get pregnant. What is the probability that a sexually active woman who uses condoms regularly will get pregnant at least once in a 5-year period?

22. The average efficacy of a birth control pill is about 97.5% per year (see Exercise 20) and the average efficacy of a condom is 86% per year (see Exercise 21). If a sexually active woman uses both the pill and condoms (and assuming that the two preventative measures are independent), what is the probability that she will get pregnant at least once in 5 years?

23. Consider the number of conditions we need to check to prove independence. To show that three events are independent, we need to check four conditions (see the text following Example 5.2).

    (a) List all the conditions we need to check to prove that the events A, B, C, and D are independent.

    (b) How many conditions do we need to check to prove that five events are independent?

    (c) Once you learn about counting (Section 10), show that, in order to prove that n events are independent, we have to check $\binom{n}{2} + \binom{n}{3} + \cdots + \binom{n}{n-1} + \binom{n}{n}$ conditions.

24. Related to Example 5.6: assume that A, B, and C are events in S, and that B and C form a partition of S. We will prove that $P(A\,|\,B) = P(A\,|\,C)$ if and only if A and B and A and C are independent.

    (a) Assume that A and B and A and C are independent. Using the definitions of the independence of two events and of the conditional probability, show that $P(A\,|\,B) = P(A\,|\,C)$.

    (b) Rewrite the condition $P(A\,|\,B) = P(A\,|\,C)$ using the definition of conditional probability.

    (c) Explain why $P(B) + P(C) = 1$. Explain why $P(A) = P(A \cap B) + P(A \cap C)$.

    (d) Use (c) to eliminate the terms involving C from the equation you obtained in (b), and then simplify to show that $P(A \cap B) = P(A)P(B)$.

    (e) Explain how (a) and (d) complete the proof.

    (f) Using what you just proved, explain why $P(V_{t+1}\,|\,V_t) = 0.75$ and $P(V_{t+1}\,|\,N_t) = 0.2$ imply that V_{t+1} and V_t are not independent.

25. We verify the formulas that were used in Example 5.7.

    (a) Assume that $g_{t+1} = ag_t$, where $a > 0$. Given the value g_0, find g_1, g_2, and g_3, and then find the general formula for g_t. Apply the formula you obtained, with $a = 1 - m$, to get the formula for g_t in (5.3).

    (b) Show that $r_t = (r_0 - 1)(1 - m)^t + 1$ satisfies $r_{t+1} = (1 - m)r_t + m$. (The formula for r_t is not as easy to derive as the one in (a).)

<table>
<tr><td>6</td><td>Discrete Random Variables</td></tr>
</table>

In many cases, the outcomes of experiments involving chance are real numbers. If this is the case, we use **random variables** to describe the outcomes.

Definition 15 Random Variable

Assume that S is a sample space of a random experiment. A *random variable X* is a function from S into the set of real numbers.

Thus, a random variable assigns a unique real number to each event in S. To denote a random variable, we write

$$X: S \to \mathbb{R}$$

Unlike the functional notation in calculus, we use uppercase letters (X, Y, Z, and so on) to identify random variables.

A set is called *finite* if it contains finitely many elements (or none at all); otherwise, it is an *infinite set*. An infinite set is called *countable* if all of its elements can be listed in a sequence; for instance, the set of positive integers $\{1, 2, 3, \ldots\}$ is countable, and so is the set of all odd numbers. It is not possible to list all real numbers that belong to the interval $[0, 1]$ in a sequence, and therefore $[0, 1]$ is not a countable set. Infinite sets that are not countable are called *uncountable*.

Definition 16 Discrete and Continuous Random Variables

If the range of a random variable X is a finite or a countable set, then X is called a *discrete random variable*. Otherwise, if its range is an uncountable set, a random variable X is called a *continuous random variable*.

We focus on discrete random variables first. Our study of continuous random variables begins in Section 13.

Discrete Random Variables

We start by investigating examples of discrete random variables that we will extensively use to illustrate concepts that we define in this and in the forthcoming sections.

Example 6.1 Tossing a Coin Three Times in a Row

Consider a random experiment that consists of tossing a fair coin three times in a row and recording the side that comes up. The sample space is the set

$$S = \{\text{HHH}, \text{THH}, \text{HTH}, \text{HHT}, \text{TTH}, \text{THT}, \text{HTT}, \text{TTT}\}$$

When defining a random variable—as for any function—we need to specify the rule of assignment.

Define X to be a random variable that counts the number of tails that show up. The values of X are

$$X(\text{HHH}) = 0, \; X(\text{THH}) = 1, \; X(\text{HTH}) = 1, \; X(\text{THT}) = 2, \; X(\text{TTT}) = 3,$$

and so on.

We define a few more random variables: let the random variable Y count how many heads in a row show up. Define Z as follows: if no heads show up, then $Z = 0$; otherwise, Z is the first occurrence of H.

Assume that T is worth 3 points and H is worth -2 points. The random variable W calculates the net worth of the three tosses.

The values of the four random variables are given in Table 6.1. They are all discrete random variables, since their ranges are finite sets.

Table 6.1

event	X	Y	Z	W
HHH	0	3	1	-6
THH	1	2	2	-1
HTH	1	1	1	-1
HHT	1	2	1	-1
TTH	2	1	3	4
THT	2	1	2	4
HTT	2	1	1	4
TTT	3	0	0	9

Example 6.2   **Genetics**

Consider the sample space $S = \{RR, RG, GG\}$ of genotypes involving a dominant allele R (red eyes) and a recessive allele G (green eyes). Define the random variable Y as follows:

$$Y = \begin{cases} 1 & \text{if the combination yields red eyes} \\ 0 & \text{if the combination yields green eyes} \end{cases}$$

Then $Y(RR) = 1$, $Y(RG) = 1$, and $Y(GG) = 0$. In words, Y is a discrete random variable whose range is the two-element set $\{0, 1\}$.

Example 6.3   **Dynamics of Disappearance and Recurrence of a Virus**

Recall the situation of Example 5.6: if, at some time, a virus is present in a population, then it will be present the following month with a probability of 0.75 (thus, it will disappear with probability 0.25). If the virus is absent from the population, then it will be absent the following month with probability 0.8 (i.e., it will (re)appear within the population with probability 0.2).

Assume that, currently, the virus is absent from the population. Define X to be a random variable that counts the number of virus-free months in the next 3-month period.

Denoting the presence of a virus by V and its absence by N, we write the sample space as

$$S = \{VVV, NVV, VNV, VVN, NNV, NVN, VNN, NNN\}$$

For instance, NVV describes the event that the population (which starts virus-free, by assumption) is virus-free for a month; then the virus appears and remains in the population for two months.

Thus, $X(VVV) = 0$, $X(NVN) = 2$, $X(NVV) = 1$, and so on. X is a discrete variable with (finite) range $\{0, 1, 2, 3\}$.

How can the range of a random variable be an infinite countable set? Look at the following example.

Example 6.4   Random Motion of a Molecule

Recall the random motion of a single molecule that we studied in Example 2.6: initially, a molecule is inside a given region. During each time interval, it remanis inside the region with probability 0.85 and leaves the region with probability 0.15. Once it leaves the region, the molecule does not return.

Define the random variable X to be the time interval during which the molecule leaves the region. Using the symbol I to denote that the molecule is inside the region and O that it is outside, the sample space can be written as

$$S = \{O,\ IO,\ IIO,\ IIIO, \dots\}$$

(for instance, the event IIO descibes the situation where the molecule remains within the region during the first two time intervals and leaves the region during the third time interval). The values of X are

$$X(O) = 1,\ X(IO) = 2,\ X(IIO) = 3,\ X(IIIO) = 4, \dots$$

Since the molecule can leave during any time interval (could be the fifth, could be the millionth), the range of X consists of all positive integers. It follows that X is a discrete random variable whose range is an infinite countable set.

Later, we will need to know the probabilities of events in S. It is given that $P(I) = 0.85$ and $P(O) = 0.15$. Note that the probabilities that the molecule leaves the region (or stays inside) remain unchanged throughout the whole experiment. This means that the behaviour of the molecule during any time interval is independent of what happened in the past. Using the independence of events,

$P(IO) = P(\text{I during the first time interval and O during the second time interval})$

$\quad = P(\text{I during the first time interval})P(\text{O during the second time interval})$

$\quad = P(I)P(O)$

$\quad = (0.85)(0.15) = 0.1275$

Likewise,

$$P(IIIO) = P(I)P(I)P(I)P(O) = (0.85)^3(0.15) = 0.0921$$

and so on.

Example 6.5   Modified Random Walk

Consider a modification of the random walk experiment discussed in Example 2.8 in Section 2. A particle is released at the location $x = 0$ and moves, during each time interval, either left or right for one distance unit with a fifty-fifty chance. Now the modification: when the particle reaches the location $x = 3$ or $x = -3$, it is absorbed. Define

$$X = \text{time needed for a particle to be absorbed}$$

The sample space consists of all random paths that end at 3 or -3:

$$S = \{0123, 012123, 0\text{-}1\text{-}2\text{-}1\text{-}2\text{-}3, 0\text{-}1010\text{-}1\text{-}2\text{-}3, 0\text{-}101212123, \dots\}$$

For instance, 0-1-2-1-2-3 represents the particle moving from 0 to -1 (during the first time interval), then to -2 (during the second time interval), then back to -1 (third interval), then to -2 (fourth interval), and finally to -3 (fifth interval), where the particle is absorbed. Thus, $X(0\text{-}1\text{-}2\text{-}1\text{-}2\text{-}3) = 5$. Likewise, $X(0123) = 3$, $X(012123) = 5$, and $X(0\text{-}101212123) = 9$.

We conclude that the range of X is $\{3, 5, 7, 9, 11, 13, \dots\}$. (Why odd numbers only? See Exercise 31.) The fact that the range of X is an infinite countable set makes X a discrete random variable.

Next, we assign probabilities to a random variable by assigning a probability to each value that the random variable assumes.

We start with *discrete* random variables whose range is a *finite* set.

Example 6.6 Assigning Probabilities: Rolling Two Dice

Consider a random experiment of rolling two dice. The sample space S consists of 36 simple events

$$S = \{(1,1),(1,2),(1,3),(1,4),(1,5),(1,6),(2,1),\ldots,(6,5),(6,6)\}$$

(where (m,n) means that the number m came up on the first die and n came up on the second). Since all events are equally likely, the probability that any one occurs is $1/36$.

Define X to be the sum of the numbers that come up; its values are given in the second column in Table 6.2. As we can see, the range of X is $\{2,3,4,\ldots,11,12\}$, and so X is a discrete random variable with finite range.

In the third column we calculated the probabilities for X. (How? See Theorem 4 in Section 3 and Example 3.12).

Table 6.2

Events	$X=$ sum	Probability
(1,1)	2	$P(X=2) = 1/36$
(1,2), (2,1)	3	$P(X=3) = 2/36$
(1,3), (2,2), (3,1)	4	$P(X=4) = 3/36$
(1,4), (2,3), (3,2), (4,1)	5	$P(X=5) = 4/36$
(1,5), (2,4), (3,3), (4,2), (5,1)	6	$P(X=6) = 5/36$
(1,6), (2,5), (3,4), (4,3), (5,2), (6,1)	7	$P(X=7) = 6/36$
(2,6), (3,5), (4,4), (5,3), (6,2)	8	$P(X=8) = 5/36$
(3,6), (4,5), (5,4), (6,3)	9	$P(X=9) = 4/36$
(4,6), (5,5), (6,4)	10	$P(X=10) = 3/36$
(5,6), (6,5)	11	$P(X=11) = 2/36$
(6,6)	12	$P(X=12) = 1/36$

Example 6.7 Assigning Probabilities: Tossing a Coin

Consider the experiment of tossing a coin three times in a row (Example 6.1). Assuming that all eight outcomes are equally likely (which is guaranteed if the coin is fair), the probability of each outcome occurring is $1/8$. The random variable X counts the number of tails. The probability that X is equal to 0 is

$$P(X=0) = P(\text{HHH}) = \frac{1}{8}$$

Looking at Table 6.1, we see that there are three events whose outcome contains one T; thus,

$$P(X=1) = P(\text{THH}, \text{HTH}, \text{HHT}) = \frac{3}{8}$$

Likewise, $P(X=2) = 3/8$ and $P(X=3) = 1/8$.

We display these probabilities in a table (Table 6.3) and as a histogram (Figure 6.1). The horizontal axis in a histogram contains the values of the random variable and the vertical axis shows the probabilities.

Table 6.3

x	$P(X=x)$
0	1/8
1	3/8
2	3/8
3	1/8

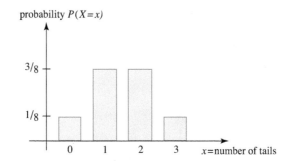

FIGURE 6.1

The histogram for the random variable X

In Exercise 11 we assign probabilities to the remaining random variables from Example 6.1.

Example 6.8 **Assigning Probabilities to the Random Variable in Example 6.3**

In order to assign probabilities to X we need to find the probability of each event. Unlike the coin-tossing experiment, the outcomes in this case are not equally likely.

Note that the probabilities remain unchanged through time. This means that the present behaviour of the virus is unaffected by (independent of) its past. Keep in mind that the initial state is N (virus not present). Using independence, we compute (the probabilities are given in Example 6.3)

$$P(\text{VVV}) = P(\text{V during the first month and V during the second month}$$
$$\text{and V during the third month})$$
$$= P(\text{V during the first month})P(\text{V during the second month})$$
$$P(\text{V during the third month})$$
$$= (0.2)(0.75)(0.75) = 0.1125$$

Likewise,

$$P(\text{VNV}) = P(\text{V during the first month})P(\text{N during the second month})$$
$$P(\text{V during the third month})$$
$$= (0.2)(0.25)(0.2) = 0.01$$

and

$$P(\text{NVV}) = P(\text{N})P(\text{V})P(\text{V }) = (0.8)(0.2)(0.75) = 0.12$$

In Table 6.4 we show all events in the sample space, their probabilities, and the values of the random variable X.

Table 6.4

Event	Probability	Value of X
VVV	$(0.2)(0.75)(0.75) = 0.1125$	0
NVV	$(0.8)(0.2)(0.75) = 0.12$	1
VNV	$(0.2)(0.25)(0.2) = 0.01$	1
VVN	$(0.2)(0.75)(0.25) = 0.0375$	1
NNV	$(0.8)(0.8)(0.2) = 0.128$	2
NVN	$(0.8)(0.2)(0.25) = 0.04$	2
VNN	$(0.2)(0.25)(0.8) = 0.04$	2
NNN	$(0.8)(0.8)(0.8) = 0.512$	3

Thus, $P(X = 0) = 0.1125$,

$$P(X = 1) = 0.12 + 0.01 + 0.0375 = 0.1675$$

and so on; see Table 6.5.

Table 6.5

x	$P(X = x)$
0	0.1125
1	0.1675
2	0.208
3	0.512

Example 6.9   Assigning Probabilities: Random Walk

Figure 6.2 shows a computer-generated histogram representing the probabilities for the position of a particle after five steps of random motion starting at $x = 0$.

Recall that the random variable X records the location of the particle. Thus, $P(X = x)$ is the probability that the particle is located at x after five steps. The histogram is based on repeating the experiment 1,000 times.

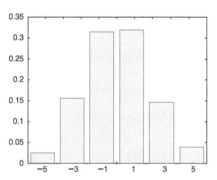

FIGURE 6.2

Histogram of the random walk experiment

In Table 6.6 we compare the simulated probabilities from Figure 6.2 against the true probabilities (for the calculation of the true probabilities, see Exercise 30; all we have to do is to count the number of different paths by which a particle can reach a certain location).

Table 6.6

Value of X	Simulated probability	True probability
$X = -5$	0.025	$1/32 \approx 0.03125$
$X = -3$	0.156	$5/32 \approx 0.15625$
$X = -1$	0.315	$10/32 \approx 0.3125$
$X = 1$	0.319	$10/32 \approx 0.3125$
$X = 3$	0.146	$5/32 \approx 0.15625$
$X = 5$	0.039	$1/32 \approx 0.03125$

Probability Mass Function and Cumulative Probability Function

We now associate a couple of important functions with a random variable.

Definition 17 Probability Mass Function

Let X be a random variable. The function

$$p(x) = P(X = x)$$

is called the *probability mass function*. The function $p(x)$ is said to describe a *probability distribution* of X.

Note that the domain of $p(x)$ is the range of the random variable X. Because p is defined as a probability, it follows that $0 \leq p(x) \leq 1$ for all x. As well, the sum of all values of $p(x)$ is 1. In symbols,

$$\sum_x p(x) = 1 \quad \text{or} \quad \sum_x P(X = x) = 1$$

where the sum is calculated over all x (that's why the x is under the summation sign) for which $P(X = x) > 0$. (This last part of the sentence means that we will be able to avoid calculating infinite sums.)

Convince yourself that the sum $\sum_x p(x)$ is 1 by adding up all probabilities in Tables 6.2, 6.3, and 6.5, and the true probabilities in Table 6.6.

One way to describe how the probabilities are assigned to a random variable X is to define the probability mass function $p(x)$. Now we introduce another function that describes the probability distribution of X.

Definition 18 Cumulative Distribution Function

Let X be a random variable. The *cumulative distribution function* $F(x)$ of X is defined as

$$F(x) = P(X \leq x)$$

We often abbreviate cumulative distribution function by c.d.f.

The domain of $F(x)$ consists of all real numbers. Since $F(x)$ is defined as a probability, $0 \leq F(x) \leq 1$ for all x.

The probability mass function and the cumulative distribution function are equivalent: if we know one, we can easily figure out the other one.

Example 6.10 Finding the Cumulative Distribution Function from the Probability Mass Function

Consider the probability mass function for the random variable X that we obtained in Example 6.8; see Table 6.5. Find a formula for the cumulative distribution function $F(x)$ and sketch its graph.

▶ To get a feel for $F(x)$, we calculate a few values. Since X does not assume negative values, the event $\{X \leq -3\}$ is empty. Thus

$$F(-3) = P(X \leq -3) = P(\emptyset) = 0$$

Likewise, $F(x) = 0$ for all $x < 0$. When $x = 0$,

$$F(0) = P(X \leq 0) = P(X = 0) = 0.1125$$

and

$$F(0.99) = P(X \leq 0.99) = P(X = 0) = 0.1125$$

In the same way we calculate other values for F; for instance,
$$F(1) = P(X \le 1) = P(X = 0) + P(X = 1) = 0.1125 + 0.1675 = 0.28$$
(we used the fact that the events $\{X = 0\}$ and $\{X = 1\}$ are mutually exclusive; thus, the probability of their union $\{X \le 1\} = \{X = 0\} \cup \{X = 1\}$ is the sum of the probabilities). As well,
$$F(1.7) = P(X \le 1.7) = P(X = 0) + P(X = 1) = 0.1125 + 0.1675 = 0.28$$
and
$$F(2) = P(X \le 2) = P(X = 0) + P(X = 1) + P(X = 2)$$
$$= 0.1125 + 0.1675 + 0.208 = 0.488$$
If $x \ge 3$, then
$$F(x) = P(X \le x) = P(X = 0) + P(X = 1) + P(X = 2) + P(X = 3)$$
$$= 0.1125 + 0.1675 + 0.208 + 0.512 = 1$$
What does the graph of F look like?

Going from left to right, the values of F are zero until x reaches the smallest value in the range of X (which is $x = 0$). Then, the graph of F jumps by the amount $P(X = 0)$ and remains constant until it encounters the next value in the range of X, namely $x = 1$. There, it jumps by the value of $P(X = 1)$. So, at $x = 1$, the value of F is equal to the sum of the two jumps. The pattern continues — the graph of F is a piecewise constant function. We find
$$F(x) = \begin{cases} 0 & x < 0 \\ 0.1125 & 0 \le x < 1 \\ 0.28 & 1 \le x < 2 \\ 0.488 & 2 \le x < 3 \\ 1 & x \ge 3 \end{cases}$$
The graph of F is given in Figure 6.3 (recall that filled circles represent the values of F).

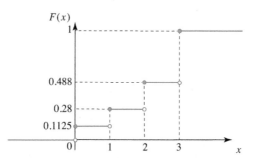

FIGURE 6.3

The graph of $F(x)$

The graph of $F(x)$ is piecewise constant, with discontinuities at $x = 0$, 1, 2, and 3. Note that the left and the right limits of F at these values of x are not equal. For instance,
$$\lim_{x \to 2^+} F(x) = 0.488 \quad \text{whereas} \quad \lim_{x \to 2^-} F(x) = 0.28$$
More precisely, because the *right* limit of $F(x)$ as x approaches 0, 1, 2, or 3 is equal to the value of F, it follows that $F(x)$ is right-continuous at 0, 1, 2, and 3.

At every discontinuity, a positive number is added to F; thus, F is a non-decreasing function (we cannot say "increasing" because F is not increasing on its flat (horizontal) parts). Moreover,
$$\lim_{x \to -\infty} F(x) = 0 \quad \text{and} \quad \lim_{x \to \infty} F(x) = 1$$

To summarize:

Let X be a discrete random variable with probability mass function $p(x) = P(X = x)$. The cumulative distribution function $F(x)$ satisfies the following:

(a) $0 \leq F(x) \leq 1$ for all $x \in \mathbb{R}$.

(b) $F(x)$ is non-decreasing (i.e., constant, or increasing) for all $x \in \mathbb{R}$.

(c) $F(x)$ has jump discontinuities of size $p(x)$ at those x where $p(x) > 0$ (i.e., at all x in the range of X where $P(X = x) > 0$).

(d) $F(x)$ is right-continuous at all points x where $p(x) > 0$.

(e) $\lim_{x \to -\infty} F(x) = 0$ and $\lim_{x \to \infty} F(x) = 1$.

Now reverse the situation: given a cumulative distribution function, we recover the probability distribution.

Example 6.11   Finding the Probability Mass Function from the Cumulative Distribution Function

The following is the cumulative distribution function of a random variable X:

$$F(x) = \begin{cases} 0 & x < -10 \\ 0.2 & -10 \leq x < -2.2 \\ 0.25 & -2.2 \leq x < 3 \\ 0.8 & 3 \leq x < 4 \\ 1 & x \geq 4 \end{cases}$$

Find the probability mass function of X.

▶ First of all, note that $F(x)$ satisfies the conditions (a) to (e) that we listed in the summary above, so it could indeed be the cumulative distribution function of some random variable X. (For instance, if the values 0.25 and 0.8 in the definition of F were switched, or the value 1 replaced by 1.3, or the condition $-10 \leq x < -2.2$ replaced by $-10 < x \leq -2.2$, then F could no longer be a cumulative distribution function of a random variable.)

We look at the discontinuities: $x = -10, -2.2, 3$, and 4. The sizes of the jumps are equal to the non-zero probabilities (i.e., the non-zero values of $p(x)$). Thus,

$$p(-10) = 0.2$$
$$p(-2.2) = 0.25 - 0.2 = 0.05$$
$$p(3) = 0.8 - 0.25 = 0.55$$
$$p(4) = 1 - 0.8 = 0.2$$

are the only non-zero values of p; see Table 6.7.

The sum is equal to

$$p(-10) + p(-2.2) + p(3) + p(4) = 0.2 + 0.05 + 0.55 + 0.2 = 1$$

which means that, indeed, there could not be any other values for x where $p(x) = P(X = x) > 0$.   ▲

Table 6.7

x	$P(X = x)$
-10	0.2
-2.2	0.05
3	0.55
4	0.2

Example 6.12   Finding a Cumulative Distribution Function

Consider the experiment of rolling two dice from Example 6.6. The random variable X is the sum of the numbers that come up, and denote by $F(x)$ its cumulative distribution function. Using the probabilities given in Table 6.2, find $F(5)$, $F(5.99)$, $F(6)$, and $F(6.01)$.

▶ Using the definition of the cumulative distribution function,

$$F(5) = P(X \leq 5)$$
$$= P(X = 2) + P(X = 3) + P(X = 4) + P(X = 5)$$
$$= \frac{1}{36} + \frac{2}{36} + \frac{3}{36} + \frac{4}{36} = \frac{10}{36}$$

(note that we used the mutual exclusivity of the events).

$F(x)$ is constant between the consecutive values in the range of X. After 5, the next value in the range is 6, and so $F(5.99) = F(5) = 10/36$. When x reaches 6, F jumps by $P(X = 6) = 5/36$. Thus,

$$F(6) = F(5) + P(X = 6) = \frac{10}{36} + \frac{5}{36} = \frac{15}{36}$$

Because F is constant between 6 and 7, $F(6.01) = F(6) = 15/36$.   ◢

Summary   A **random variable** is a function from a sample space into the set of real numbers. If its range is finite or countable, then it is a **discrete random variable.** The range of a **continuous random variable** is an uncountable set. There are two ways to assign probabilities to a discrete random variable: we can define the **probability mass function** or, equivalently, the **cumulative distribution function.** The probability mass function of a finite discrete variable has finitely many non-zero values that add up to 1. The cumulative distribution function gives the probability that the random variable is smaller than or equal to a particular value.

6   Exercises

1. Let X count the number of tosses until we toss five heads in a row. Is X a discrete or a continuous random variable? Is it finite or infinite?

2. Let X be the location of a particle after 100 steps of the random walk (see Example 2.8 in Section 2). Is X a discrete or a continuous random variable? Is it finite or infinite?

3. Assume that the range of a random variable X is the set $\{1, 2, 3\}$. Define $p(1) = 0.16$, $p(2) = 0.54$, and $p(3) = 0.29$. Can p be a probability mass function of X? Why or why not?

4. Assume that the range of a random variable X is the set $\{0, 1, 2, 3\}$. Define $p(0) = 0$, $p(1) = 0.16$, $p(2) = 0.54$, and $p(3) = 0.3$. Can p be a probability mass function of X? Why or why not?

5. Explain why

$$F(x) = \begin{cases} 0 & x < 0 \\ 0.32 & 0 \leq x < 1 \\ 0.31 & 1 \leq x < 2 \\ 1 & x \geq 2 \end{cases}$$

cannot be a cumulative distribution function of a random variable.

6. Explain why

$$F(x) = \begin{cases} 0.1 & x < 0 \\ 0.2 & 0 \leq x < 1 \\ 0.6 & 1 \leq x < 2 \\ 1 & x \geq 2 \end{cases}$$

cannot be a cumulative distribution function of a random variable.

▽ 7–10 ▪ Find the probability mass function of each random variable X.

7. Toss a coin four times. The random variable X counts the number of trials until the first tails. If all four tosses show heads, then define $X = 0$.

8. Toss a coin four times. The random variable X counts the number of heads.

9. Roll two dice. X is the maximum of the numbers that show up.

◣ 10. Roll two dice. X is the absolute value of the difference of the numbers that show up.

11. Based on Table 6.1, find the probability distributions for the random variables Y, Z, and W from Example 6.1. Make sure that the probabilities add up to 1.

12. If, at some time, a virus is present in a population, then it will be present the following month with probability 0.75 (thus, it will disappear with probability 0.25). If the virus is absent from the population, then it will be absent the following month with probability 0.8 (i.e., it will (re)appear within the population with probability 0.2). Assume that at this moment the virus is present in the population. Find the probability mass function for the random variable $X =$ "number of virus-free months in the 2-month period from now."

13. If, at some time, a virus is present in a population, then it will be present the following month with probability 0.4 (thus, it will disappear with probability 0.6). If the virus is absent from the population, then it will be absent the following month with probability 0.7 (i.e., it will (re)appear within the population with probability 0.3). Assume that at this moment the virus is absent from the population. Find the probability mass function for the random variable $X =$ "number of virus-free months in the 2-month period from now."

14. A couple of bonobo monkeys have a baby monkey each year: a female with probability 0.55, and a male with probability 0.45. Let $B =$ "number of female baby monkeys born to the couple in a 3-year period." Find the probability mass function for B.

15. A couple of rhesus monkeys have a baby monkey each year with a chance of 65% that the baby will be dark brown and 35% that it will be light brown and grey. Let $R =$ "number of dark brown baby monkeys born to the couple in a 3-year period." Find the probability mass function for R.

16. A couple of rhesus monkeys have a baby monkey each year with a chance of 25% that the baby will be dark brown, 35% that it will be light brown, and 40% that it will be grey. Let $R =$ "number of grey baby monkeys born to the couple in a 2-year period." Find the probability mass function for R.

17. A couple of bonobo monkeys have a baby monkey each year with a chance of 15% that the baby will have red eyes, 5% that it will have blue eyes, and 80% that it will have brown eyes. Let $B =$ "number of blue-eyed baby monkeys born to the couple in a 2-year period." Find the probability mass function for B.

▽ 18–21 ▪ Draw a histogram for the following probability mass functions, and a pick the word or phrase among "symmetric," "skewed left," "skewed right," and "uniform" that best describes it.

18. $p(1) = 0.35$, $p(2) = 0.2$, $p(3) = 0.1$, $p(4) = 0.15$, $p(5) = 0.1$, $p(6) = 0.05$, $p(7) = 0.04$, $p(8) = 0.01$

19. $p(1) = 0.15$, $p(2) = 0.1$, $p(3) = 0.15$, $p(4) = 0.15$, $p(5) = 0.1$, $p(6) = 0.1$, $p(7) = 0.15$, $p(8) = 0.1$

20. $p(1) = 0$, $p(2) = 0.2$, $p(3) = 0.1$, $p(4) = 0.2$, $p(5) = 0.15$, $p(6) = 0.1$, $p(7) = 0.25$, $p(8) = 0$

◣ 21. $p(1) = 0$, $p(2) = 0$, $p(3) = 0.1$, $p(4) = 0$, $p(5) = 0.1$, $p(6) = 0.3$, $p(7) = 0.5$, $p(8) = 0$

22–25 ▪ Given is the cumulative distribution function of a random variable X. Find the probability mass function of X.

22.
$$F(x) = \begin{cases} 0 & x < 3 \\ 0.1 & 3 \le x < 5 \\ 0.4 & 5 \le x < 10 \\ 1 & x \ge 10 \end{cases}$$

23.
$$F(x) = \begin{cases} 0 & x < 0.7 \\ 0.3 & 0.7 \le x < 1 \\ 0.7 & 1 \le x < 1.2 \\ 1 & x \ge 1.2 \end{cases}$$

24.
$$F(x) = \begin{cases} 0 & x < -4 \\ 0.5 & -4 \le x < -2 \\ 0.65 & -2 \le x < -1 \\ 0.95 & -1 \le x < 0 \\ 1 & x \ge 0 \end{cases}$$

25.
$$F(x) = \begin{cases} 0 & x < 1/2 \\ 0.1 & 1/2 \le x < 1 \\ 0.5 & 1 \le x < 3/2 \\ 0.8 & 3/2 \le x < 3 \\ 1 & x \ge 3 \end{cases}$$

26–29 ▪ Given is the probability mass function of a discrete random variable X. Find the cumulative distribution function of X and sketch its graph.

26.

x	$P(X = x)$
-2	0.15
1	0.15
3	0.45
4	0.25

27.

x	$P(X = x)$
0	0.25
1	0.25
2	0.25
3	0.25

28.

x	$P(X = x)$
-2	0.25
-1	0.1
0	0.15
1	0.2
2	0.3

29.

x	$P(X = x)$
0	0.8
1	0.05
2	0.05
3	0.05
4	0.05

30. Continue the diagram in Example 3.10 in Section 3 until you complete the fifth step of the random walk. The random variable X records the location after five steps of random motion.

   (a) Explain why there is only one way to reach $x = 5$. List all routes that end at $x = 3$.

   (b) List all ten routes in which a random walk ends at $x = 1$ after five steps.

   (c) Explain why the fifth step in the random walk contains 32 different routes (i.e., 32 different ways to start at $x = 0$ and end at one of -5, -3, -1, 1, 3, and 5 after five steps). Use this fact, and your answers to (a) and (b), to build a probability distribution for X.

31. Assume that 0 is an even number, and classify negative numbers by looking at their absolute values (-3 is odd, -4 is even, etc.).

   (a) Consider the usual random walk that we introduced in Example 2.8. List all possible locations of a particle after 1, 2, 3, 4, and 5 steps of a random walk. You will notice that, after an even (odd) number of steps, the particle is located at even- (odd-) numbered locations. Explain why this is true.

(b) Consider the modified random walk from Example 6.5. Using (a), explain why the range of the random variable X consists of odd numbers only.

32. Given is the histogram of a random variable X. Find its probability mass function and cumulative distribution function.

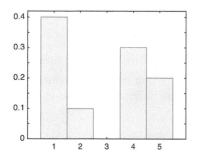

33. Given is the histogram of a random variable X. Find its probability mass function and cumulative distribution function.

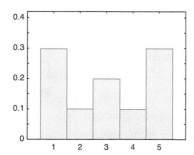

7	The Mean, the Median, and the Mode

A random variable is fully described by its probability mass function or, equivalently, by its cumulative distribution function. In reality, there are situations where it is not practical (or useful) to use these functions to represent a random variable. Instead, we extract certain information (a single number, or a small collection of numbers) that, in a satisfactory way, represents the random variable. For instance, a weather report would be incomprehensible (and useless) if it consisted of a long list of measurements of air temperature taken at regular intervals throughout the day. Reporting the **average** daily temperature instead makes a lot more sense.

In this section we discuss ways in which we can compute information about the **centre of the distribution.** In the following section we introduce the measurements of the spread of data.

Expected Value

We start with an example.

Example 7.1   Daily Milk Production on Milky Way Farm

The Milky Way Farm in Eastern Quebec has 30 cows. Its owner recorded the daily production D of milk, in litres:

$$30, 18, 20, 27, 30, 20, 18, 29, 30, 30, 27, 22, 18, 30, 20,$$
$$24, 18, 27, 22, 29, 29, 20, 29, 18, 18, 30, 27, 20, 30, 25 \tag{7.1}$$

How much milk does a Milky Way Farm cow produce, on average, in a day?

The answer is straightforward: we add up the amount of milk each cow produces in a day and divide by 30. Denoting the average daily production by $\overline{D}$, we compute

$$\overline{D} = \frac{1}{30}(30 + 18 + 20 + 27 + 30 + 20 + 18 + \cdots + 30 + 25) = \frac{735}{30} = 24.5$$

litres. The number $\overline{D}$ is called the *average value* or the *mean.* (more precisely, it is called the *arithmetic mean*).

Adding up 30 numbers is not fun. To make this calculation more pleasant, we organize the production numbers into groups (see Table 7.1).

Table 7.1

Production	18	20	22	24	25	27	29	30
Frequency	6	5	2	1	1	4	4	7

We recorded the *frequencies,* i.e., we counted how many times each number occurs in the list (7.1); for instance, six cows produced exactly 18 L of milk, five cows produced 20 L of milk, and so on. Now we compute the average based on the frequencies:

$$\overline{D} = \frac{1}{30}(6 \cdot 18 + 5 \cdot 20 + 2 \cdot 22 + 1 \cdot 24 + 1 \cdot 25 + 4 \cdot 27 + 4 \cdot 29 + 7 \cdot 30)$$
$$= \frac{735}{30} = 24.5 \tag{7.2}$$

We further rearrange the terms in (7.2)

$$\overline{D} = \frac{6}{30} \cdot 18 + \frac{5}{30} \cdot 20 + \frac{2}{30} \cdot 22 + \frac{1}{30} \cdot 24 + \frac{1}{30} \cdot 25 + \frac{4}{30} \cdot 27 + \frac{4}{30} \cdot 29 + \frac{7}{30} \cdot 30$$
$$= \frac{735}{30} = 24.5 \tag{7.3}$$

and redo Table 7.1 using *relative frequencies;* see Table 7.2.

Table 7.2

Production	18	20	22	24	25	27	29	30
Relative frequency	6/30	5/30	2/30	1/30	1/30	4/30	4/30	7/30

The relative frequency information associates the amount of milk with the *ratio* of cows that produce it. For instance, $6/30 = 20\%$ of the cows give 18 L of milk per day each; $5/30 = 1/6$ of the cows produce 20 L of milk each, and so on.

Note that the *relative frequencies are actually probabilities,* based on the sample space S consisting of the 30 cows on the farm. So Table 7.2 represents the probability distribution for the random variable $D = $ "daily production of milk" defined on S. We interpret the columns in Table 7.2 as $P(D = 18) = 6/30$, $P(D = 20) = 5/30$, and so on.

The average $\overline{D}$ in (7.3) is the sum of terms, each of the form

(ratio of cows that give d litres of milk) $\cdot$ (d litres of milk)

i.e.,

(probability that $D = d$) $\cdot d$

In symbols,

$$\overline{D} = \sum_d P(D = d) \cdot d$$

where the sum is calculated over all d for which $P(D = d) > 0$.

Keeping this in mind, we now define the *mean* or the *expected value* of a random variable. We abandon using the phrase "average value" to avoid the confusion that could arise from the variety of meanings associated with it.

The mean is one of the most important statistics related to a random variable.

Definition 19 Mean or Expected Value

Let X be a discrete random variable. The *mean* or the *expected value* of X is the number

$$E(X) = \sum_x x P(X = x) = \sum_x x p(x)$$

where the sum goes over all values x for which $p(x) = P(X = x)$ is not zero. ◢

When the range of X is a finite set, the sum in Definition 19 is a finite sum. Otherwise, we need to add infinitely many terms, which forces us to consider infinite series. The issues of convergence of infinite series place this case beyond the scope of this book, so we will not deal with it.

Example 7.2 Calculating Expected Value: Rolling Dice

Assume that we roll two dice and add up the numbers that come up. What is the expected value of the sum?

▶ In Example 6.6 in Section 6 we described the sample space for this experiment. The random variable X adds up the numbers that come up; thus

$$X: S \to \{2, 3, 4, 5, \ldots, 12\}$$

In Table 6.2 we find the probabilities that we need. We compute

$$E(X) = \sum_{2}^{12} x \cdot P(X = x)$$

$$= 2 \cdot P(X = 2) + 3 \cdot P(X = 3) + 4 \cdot P(X = 4) + 5 \cdot P(X = 5)$$

$$+ \cdots + 12 \cdot P(X = 12)$$

$$= 2 \cdot \frac{1}{36} + 3 \cdot \frac{2}{36} + 4 \cdot \frac{3}{36} + 5 \cdot \frac{4}{36} + 6 \cdot \frac{5}{36} + 7 \cdot \frac{6}{36} + 8 \cdot \frac{5}{36}$$

$$+ 9 \cdot \frac{4}{36} + 10 \cdot \frac{3}{36} + 11 \cdot \frac{2}{36} + 12 \cdot \frac{1}{36}$$

$$= \frac{252}{36} = 7$$

Thus, the expected value of the sum showing up on the two dice is 7. ⬛

Example 7.3   Calculating Expected Value: Virus Dynamics

In Example 6.8 in Section 6 we studied the dynamics of the appearance and disappearance of a virus and arrived at the probability distribution (repeated, for convenience, in Table 7.3) for the random variable $X = $ "number of virus-free months in a 3-month interval."

Table 7.3

x	$P(X = x)$
0	0.1125
1	0.1675
2	0.208
3	0.512

The mean of X is

$$E(X) = \sum_{0}^{3} x \cdot P(X = x)$$

$$= 0 \cdot P(X = 0) + 1 \cdot P(X = 1) + 2 \cdot P(X = 2) + 3 \cdot P(X = 3)$$

$$= 0 \cdot 0.1125 + 1 \cdot 0.1675 + 2 \cdot 0.208 + 3 \cdot 0.512$$

$$= 2.1195$$

Thus, on average, we expect to see about 2.1 virus-free months within a 3-month period (assuming no virus is present initially). ⬛

Example 7.4   Expected Value for the Random Walk

The probability distribution of the random variable $X = $ "position of a particle after five steps of random motion starting at $x = 0$" is shown in Table 7.4 (taken from Example 6.9 in Section 6).

Table 7.4

x	$P(X = x)$
-5	$1/32$
-3	$5/32$
-1	$10/32$
1	$10/32$
3	$5/32$
5	$1/32$

The expected (mean) location is

$$E(X) = \sum_{x} x \cdot P(X = x)$$

where x is in $\{-5, -3, -1, 1, 3, 5\}$. Thus,

$$E(X) = (-5) \cdot P(X = -5) + (-3) \cdot P(X = -3) + (-1) \cdot P(X = -1)$$

$$+ 1 \cdot P(X = 1) + 3 \cdot P(X = 3) + 5 \cdot P(X = 5)$$

$$= (-5)\frac{1}{32} + (-3)\frac{5}{32} + (-1)\frac{10}{32} + (1)\frac{10}{32} + (3)\frac{5}{32} + (5)\frac{1}{32}$$

$$= 0$$

⬛

The fact that the mean is 0 is not at all exciting. Because a particle moves randomly, it will end up at symmetric locations (such as -3 and 3) with equal probability (look at Table 7.4). Thus, the mean location of the particle is its starting position $x = 0$.

However, it is wrong to deduce that, on average, a particle will end its random walk at $x = 0$, where it started. We need to do a different calculation to figure out, on average, how far from the origin a particle will be after five steps of random walk.

To calculate the *distance,* we need to remove the minus signs from the final locations of the particle. So, we'll square the x values of the locations and in the end calculate the square root. More precisely, we define a new random variable $X^2 = X \cdot X$ and calculate its probability mass function (shown in the second and third columns in Table 7.5).

Table 7.5

x	x^2	$P(X = x)$
-5	$(-5)^2$	$1/32$
-3	$(-3)^2$	$5/32$
-1	$(-1)^2$	$10/32$
1	$(1)^2$	$10/32$
3	$(3)^2$	$5/32$
5	$(5)^2$	$1/32$

The expected value of X^2 is

$$E(X^2) = (-5)^2 \cdot P(X = -5) + (-3)^2 \cdot P(X = -3) + (-1)^2 \cdot P(X = -1)$$
$$+ 1^2 \cdot P(X = 1) + 3^2 \cdot P(X = 3) + 5^2 \cdot P(X = 5)$$
$$= 25 \cdot \frac{1}{32} + 9 \cdot \frac{5}{32} + \frac{10}{32} + \frac{10}{32} + 9 \cdot \frac{5}{32} + 25 \cdot \frac{1}{32}$$
$$= \frac{160}{32} = 5$$

Thus, the average position of the particle after five steps of the random walk is
$$\sqrt{E(X^2)} = \sqrt{5} \approx 2.236$$

Motivated by this discussion, we now define the expected value of a *function* of a random variable X.

Definition 20   Expected Value of a Function of a Random Variable

Assume that X is a discrete random variable and that $p(x) = P(X = x)$ is its probability mass function. Let $g(x)$ be a function of x. The *expected value* of the random variable $g(X)$ is
$$E(g(X)) = \sum_x g(x)P(X = x) = \sum_x g(x)p(x)$$

where the sum goes over all values x for which $p(x)$ is not zero.

For instance, if $g(x) = \sin x$, then
$$E(g(X)) = \sum_x \sin x \, P(X = x) = \sum_x \sin x \, p(x)$$

When $g(x) = x$, the formula in Definition 20 gives the expected value $E(X)$ of X.

Example 7.5   Expected Value of Functions of a Random Variable

The probability mass function of a random variable X is given in Table 7.6. Find $E(X + 4)$ and $E(e^{-X})$.

▶ Substituting $g(x) = x + 4$ into Definition 20 we get

$$E(X + 4) = \sum_x (x + 4)P(X = x)$$

where the sum has four terms, corresponding to $x = -2$, 0, 1 and 3. Thus

$$E(X + 4) = (-2 + 4)P(X = -2) + (0 + 4)P(X = 0) + (1 + 4)P(X = 1)$$
$$+ (3 + 4)P(X = 3)$$
$$= 2 \cdot 0.1 + 4 \cdot 0.2 + 5 \cdot 0.5 + 7 \cdot 0.2$$
$$= 4.9$$

Likewise, with $g(x) = e^{-x}$,

$$E(e^{-X}) = \sum_x e^{-x} P(X = x)$$
$$= e^{-(-2)}P(X = -2) + e^0 P(X = 0) + e^{-1}P(X = 1) + e^{-2}P(X = 3)$$
$$= e^2 \cdot 0.1 + 0.2 + e^{-1} \cdot 0.5 + e^{-2} \cdot 0.2$$
$$\approx 1.1499$$

Table 7.6

x	$P(X = x)$
-2	0.1
0	0.2
1	0.5
3	0.2

Theorem 7   Properties of the Expected Value

Let X and Y be discrete random variables and a and b be real numbers. Then

(1) $E(aX) = aE(X)$

(2) $E(X + b) = E(X) + b$

(3) $X \pm Y$ is a discrete random variable and

$$E(X \pm Y) = E(X) \pm E(Y)$$

Sometimes we join properties (1) and (2) together and write

$$E(aX + b) = aE(X) + b$$

Note that if $a = 0$, then this formula implies that $E(b) = b$. The symbol b on the left side represents the random variable (actually there is nothing random about it) all of whose (finitely many) values are equal to b. Clearly, its mean must be b.

Replacing the real number b in $E(b) = b$ by $E(X)$, we obtain the formula $E[E(X)] = E(X)$ (see Exercise 30).

The properties in Theorem 7 are intuitively clear. Suppose that we calculate the mean (call it m) of finitely many numbers. If we multiply all numbers by the same number a, then to get the new mean, we multiply the old mean m by a. If we add the same number b to all numbers, the mean will change by that same number, i.e., it will be equal to $m + b$.

For practice, we now prove formulas (1) and (2). We will not prove (3).

To prove (1), we use the formula

$$E(g(X)) = \sum_x g(x)P(X = x) = \sum_x g(x)p(x)$$

from Definition 20 with $g(x) = ax$:

$$E(aX) = \sum_x ax P(X = x) = a \sum_x x P(X = x) = aE(X)$$

(as usual, the sum is taken over all x for which the probability $P(X = x)$ is not zero). Since a is constant, we were allowed to take it out of the sum.

Now let $g(x) = x + b$ and use Definition 20 again:

$$E(X + b) = \sum_x (x + b)P(X = x)$$

$$= \sum_x xP(X = x) + \sum_x bP(X = x)$$

$$= E(X) + b\sum_x P(X = x)$$

$$= E(X) + b$$

since

$$\sum_x P(X = x) = 1$$

by the basic property of the probability mass function. This proves (2).

As attractive as it may appear (reducing information about a random variable to a single number), the expected value does not tell the whole story.

In Example 7.1 we studied the daily milk production of 30 cows on the Milky Way Farm and calculated the mean to be 24.5 L per cow per day.

Example 7.6   **Daily Milk Production on Two More Farms**

Consider the daily production of milk by 30 cows on Milkshake Farm in western Quebec:

$$23, 22, 25, 26, 27, 23, 25, 26, 26, 26, 27, 25, 25, 23, 22,$$
$$24, 25, 25, 23, 25, 24, 23, 26, 26, 25, 22, 22, 23, 24, 27 \qquad (7.4)$$

and on Butterscotch Farm in Ontario:

$$20, 17, 32, 32, 32, 20, 17, 20, 32, 30, 30, 32, 32, 18, 18,$$
$$18, 18, 32, 19, 19, 18, 18, 18, 31, 31, 31, 31, 31, 19, 19 \qquad (7.5)$$

We can easily check that the average milk production on these two farms is 24.5 L per cow per day, the same as for the Milky Way Farm cows.

However, looking at the frequencies shown in Figure 7.1, we see that the three distributions differ quite a bit. Clearly, the expected value is not able to capture their differences.

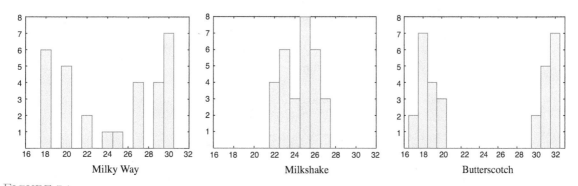

FIGURE 7.1

The distributions of milk production on the three farms

Next, we look at other statistics that describe the "centre" of a distribution, which might be able to distinguish between the three distributions.

The Median and the Mode

We now define two useful statistics for a distribution, the median and the mode. Although we can compute the median and the mode for any sample, we will see that they make sense mostly for larger samples that consist of many different values.

Definition 21 **The Median**

The *median* is the midpoint of a distribution.

Algorithm 1 below will make the meaning of the word "midpoint" precise.

Rougly speaking, the median M is a number with the property that picking a number larger than or equal to M is as likely as picking a number smaller than or equal to M.

We discuss a few examples and then explain why we used the word "roughly" in interpreting the meaning of the median.

To find the median of the set of five numbers $\{1, 7, 3, 2, 10\}$, we create the ordered list $1, 2, 3, 7, 10$ and pick the number in the centre. Thus, the median is 3; picking a number larger than or equal 3 is as likely as picking a number smaller than or equal to 3 (both probabilities are $3/5$).

Consider an even number of outcomes, say $\{1, 4, 3, 7, 2, 6\}$. In the centre of the ordered list $1, 2, 3, 4, 6, 7$ there are two numbers, 3 and 4. The median is taken to be the mean of the two numbers, i.e., $(3 + 4)/2 = 3.5$. Clearly, picking a number greater than or equal to 3.5 from the list $1, 2, 3, 4, 6, 7$ is as likely as picking a number smaller than or equal to 3.5 (both probabilities are $1/2$).

In creating the ordered list we have to make sure that we keep the frequencies. Ordering the set of data $\{3, 1, 2, 6, 3, 6, 3, 3, 11\}$, we get $1, 2, 3, 3, 3, 3, 6, 6, 11$. The central term—and hence the median—is 3. Note that in this case the probabilities are not equal; the probability of picking a number smaller than or equal to 3 is $6/9$, whereas the probability of picking a number greater than or equal to 3 is $7/9$. If we ask for strict inequalities, it still does not work: the probability of picking a number smaller than 3 is $2/9$ and the probability of picking a number greater than 3 is $3/9$.

So, the probabilities are not equal. However, if we consider a large set of data with many distinct values (that's when the median is actually useful), then, in many cases, things work. To alert the reader to this issue, we used the word "roughly" in explaining the meaning of the median.

The following algorithm makes Definition 21 fully transparent.

Algorithm 1   How to Locate the Median

To find the median of a sample set of n numbers:

(1) Create an ordered list of the numbers, from the smallest to the largest.

(2) If n is odd, then there is one number in the centre of the list, and that number is the median. To locate it, count $(n + 1)/2$ numbers from either end of the list.

(3) If n is even, then by counting $n/2$ numbers from both ends we arrive at the two centre numbers. The median is the mean of these two numbers.

Example 7.7 Calculating the Median Milk Production

Consider the milk production on the Milky Way Farm (Example 7.1). The outcomes ($n = 30$), written as an ordered list, are

$$18, 18, 18, 18, 18, 18, 20, 20, 20, 20, 20, 22, 22, 24, 25,$$
$$27, 27, 27, 27, 29, 29, 29, 29, 30, 30, 30, 30, 30, 30, 30$$

Counting $n/2 = 15$ outcomes from both sides, we arrive at 25 and 27. Thus, the median is $(25 + 27)/2 = 26$.

Here is an alternative way of thinking about the median. Table 7.7 gives the probability mass function of the random variable $X = $ "amount of milk a cow gives in a day."

Based on the data, we construct the cumulative distribution function

Table 7.7

x	$P(X = x)$
18	6/30
20	5/30
22	2/30
24	1/30
25	1/30
27	4/30
29	4/30
30	7/30

$$F(x) = \begin{cases} 0 & x < 18 \\ 6/30 & 18 \le x < 20 \\ 11/30 & 20 \le x < 22 \\ 13/30 & 22 \le x < 24 \\ 14/30 & 24 \le x < 25 \\ 15/30 & 25 \le x < 27 \\ 19/30 & 27 \le x < 29 \\ 23/30 & 29 \le x < 30 \\ 1 & x \ge 30 \end{cases}$$

To identify the median, we look for the x value that corresponds to the cumulative probability of $1/2$. We see that $F(x) = 1/2$ for $25 \le x < 27$. We take the midpoint, so the median is 26.

Consider the graph of a cumulative distribution function $F(x)$ in Figure 7.2.

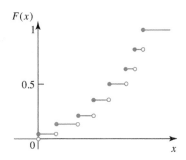

FIGURE 7.2

Locating the median from the cumulative distribution function

We would like to say that the median is the point (actually the interval, which we reduce to a point) where the cumulative distribution function crosses the horizontal line representing the probability of $1/2$. But in some cases the cumulative distribution function and the horizontal line have nothing in common. That is why we state the algorithm for finding the median in the following way.

Algorithm 2 How to Locate the Median from a Cumulative Distribution Function

Assume that X is a discrete random variable.

(1) Compute the cumulative distribution function $F(x)$ of X and sketch its graph.

(2) If the graph of $F(x)$ intersects $1/2$, then identify the interval for x where $F(x) = 1/2$. The midpoint of that interval is the median of X; see Figure 7.3a.

(3) If the graph of $F(x)$ does not intersect $1/2$, then the median of X is the mean of the values immediately above and immediately below 0.5; see Figure 7.3b.

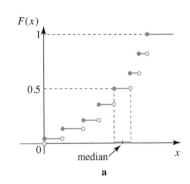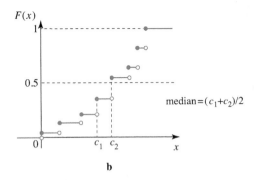

FIGURE 7.3

The median of a random variable

Definition 22   The Mode

The *mode* of a sample space is the outcome that appears most often.

The mode of the set of numbers

$$\{1, 2, 2, 3, 3, 3, 4, 4, 4, 4, 4, 5, 6, 6, 7\}$$

is 4. The mode need not be a unique number. For instance, the mode of the set

$$\{1, 2, 3, 3, 3, 4, 4, 4, 5, 5, 6, 7, 8\}$$

consists of two numbers, 3 and 4. There are situations when the mode can be calculated, but makes no sense (i.e., provides no new information about the sample). For instance, every number in the sample space

$$\{1, 1, 2, 2, 3, 3, 4, 4, 5, 5, 6, 6\}$$

is the mode.

As with the median, the mode makes (most) sense when we need to describe large sample spaces.

Example 7.8   Measures of the Centre of a Distribuiton

In Figure 7.4 we repeat the frequencies of the daily production of milk by 30 cows on each of the three farms (see Examples 7.1 and 7.6). Under each histogram, we record the mean, the median, and the mode.

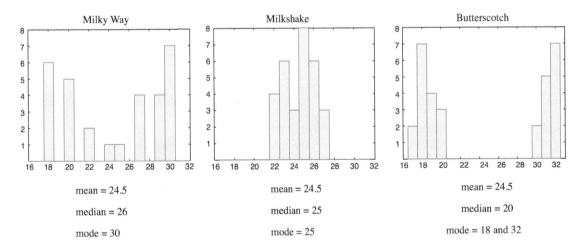

FIGURE 7.4

Comparing milk production

The distribution of the Milkshake Farm data is clustered (and somewhat symmetric), which is reflected in the fact that all three measures are equal or close to equal.

The remaning two histograms are spread out. The Milky Way data is skewed, with both the median and the mode on the same side of the mean (i.e., both are larger than the mean). The Butterscotch data has two modes, 18 and 32.

Expected Value of the Logarithm and the Geometric Mean

There are situations where the mean does not provide adequate information about the sample space, or about the behaviour of a system or a biological process.

Consider a population of fish whose per capita production rate alternates throughout the year: during a "good" season, the rate is 1.8 (say, due to abundance of food or absence of predators). During a "bad" season, the per capita production rate falls to 0.4.

The average per capita rate $(1.8 + 0.4)/2 = 1.1$ suggests that the fish population will increase on average by 10% per year.

Let's check. Assume that the initial population is $p_0 = 100,000$. We compute

$$p_1 = 1.8 \cdot 100,000 = 180,000$$
$$p_2 = 0.4 \cdot 180,000 = 72,000$$
$$p_3 = 1.8 \cdot 72,000 = 129,600$$
$$p_4 = 0.4 \cdot 129,600 = 51,840$$
$$p_5 = 1.8 \cdot 51,840 = 93,312$$
$$p_4 = 0.4 \cdot 93,312 = 37,325$$

and so on. The population is oscillating, but declining, and will eventually go extinct. If we continue our calculation, we get

$$p_{10} = 19,349$$
$$p_{20} = 3,744$$
$$p_{30} = 724$$

See Figure 7.5.

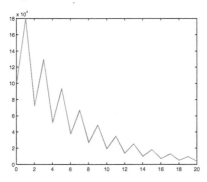

FIGURE 7.5

The fish population dynamics

This clearly contradicts the conclusion based on the mean of the per capita production rates.

The mean, as we defined it, is an adequate measurement (useful measurement) for phenomena based on *addition* such as milk production (hence the name *arithmetic mean*). However, in the case of the fish population, we *multiply* the present population by the per capita rate to get next year's population. Looking at the numbers p_1, p_2, p_3, ... again, we see that what matters is not the sum $1.8 + 0.4$ of the per capita rates, but their product $1.8 \cdot 0.4 = 0.72$. Note that

$$p_0 = 100,000$$
$$p_1 = 180,000$$
$$p_2 = 0.72 \cdot p_0 = 0.72 \cdot 100,000 = 72,000$$

$$p_3 = 0.72 \cdot p_1 = 0.72 \cdot 180{,}000 = 129{,}\ 600$$
$$p_4 = 0.72 \cdot p_2 = 0.72 \cdot 72{,}000 = 51{,}840$$
$$p_5 = 0.72 \cdot p_3 = 0.72 \cdot 129{,}600 = 93{,}312$$

and so on. The population change is obviously a multiplicative process. Let's look further into this.

Assume that a population changes according to

$$p_{t+1} = R_t p_t$$

where the per capita rate is represented by the random variable R_t. If the initial population is p_0, then

$$p_1 = R_0 p_0$$
$$p_2 = R_1 p_1 = R_1 R_0 p_0$$
$$\cdots$$
$$p_t = R_{t-1} R_{t-2} \cdots R_0 p_0 \qquad (7.6)$$

To convert (7.6) to a sum, we use logarithms:

$$\ln p_t = \ln R_{t-1} + \ln R_{t-2} + \cdots + \ln R_0 + \ln p_0 \qquad (7.7)$$

Thus, to find the logarithm of p_t we add the logarithms of the per capita rates to the logarithm of the initial population. So, if the expected value of the logarithms of R_t is positive, the sum in (7.7) will increase over time and so will the population. If the expected value is negative, the sum will decrease, and so the population will decrease as well.

So it is important to know the expected value of the logarithm of the per capita rates. In the example we started with, the rates (and thus the values of the random variable R) are 1.8 and 0.4. We calculated

$$E(R) = \frac{1}{2}(1.8) + \frac{1}{2}(0.4) = 1.1$$

The expected value of the logarithm of R is

$$E(\ln R) = \frac{1}{2}(\ln 1.8) + \frac{1}{2}(\ln 0.4) \approx -0.164$$

The expected value is negative, suggesting (correctly) a decline in the population.

Definition 23   Geometric Mean

Assume that X is a random variable such that $X > 0$ (i.e., X takes on positive values at all events in its domain). The *geometric mean* of X is the real number

$$\text{G.M.}(X) = e^{E(\ln X)}$$

Recall that the expected value of $\ln X$ is calculated as

$$E(\ln X) = \sum_x \ln x \, P(X = x) = \sum_x \ln x \cdot p(x)$$

where the sum goes over all values x for which $p(x)$ is positive (see Definition 20).

Example 7.9   Geometric Mean: Population Dynamics

Assume that the per capita production rate in a population of salmon is given by its probability mass function; see Table 7.8. Find $E(\ln R)$ and interpret the result.

Table 7.8

r	$P(R = r)$
1.8	0.25
0.7	0.65
1.2	0.1

▶ We compute

$$E(\ln R) = \sum_r \ln r \, P(R = r)$$
$$= \ln 1.8 \, P(R = 1.8) + \ln 0.7 \, P(R = 0.7) + \ln 1.2 \, P(R = 1.2)$$
$$= (\ln 1.8)(0.25) + (\ln 0.7)(0.65) + (\ln 1.2)(0.1) \approx -0.06666$$

The geometric mean

$$e^{E(\ln R)} = e^{-0.06666} \approx 0.93551$$

predicts that the population will decline at the rate of $1 - 0.93551 = 0.06449$; i.e., a bit under 6.5% per year. Figure 7.6 shows the outcomes of three simulations, all starting with initial population 1,000.

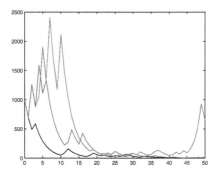

FIGURE 7.6

Three simulations of salmon population

Summary   We defined three statistics related to the **centre** of a distribution: the **mean**, the **median**, and the **mode**. The mean, or the **expected value**, corresponds to what we usually think of when we say "average value." Since distributions that are quite different can have the same mean, we need to find other statistics to describe their differences. The median is the midpoint of a distribution, and the mode is the value that appears most often. In some cases, the median and the mode do convey more information about the distribution, but they are not able to fully capture the differences among distributions (hence the next section).

7	Exercises

1. Find the median of the sets $S_1 = \{3, 2, 4, 6, 7, 10, 5\}$ and $S_2 = \{3, 2, 4, 6, 700{,}000, 1{,}000{,}000, 5\}$. What can you conclude about the median?

2. Find the mode of the sets $S_1 = \{2, 2, 2, 3, 4, 5, 4, 3\}$ and $S_2 = \{2, 2, 2, 2, 2, 1{,}000{,}000, 2{,}000{,}000\}$. What can you conclude about the mode?

3. Double each value in a data set S_1, thus obtaining the data set S_2. How are the means, the medians, and the modes of S_1 and S_2 related?

4. Add 5 to each value in a data set S_1, thus obtaining the data set S_2. How are the means, the medians, and the modes of S_1 and S_2 related?

5. A random variable X is said to be *uniformly distributed* on the set $S = \{1, 2, 3, 4, 5, 6, 7, 8, 9, 10\}$ if $P(X = k) = 1/10$ for $k = 1, 2, \ldots, 10$. What is the mean of X?

6. A random variable X is said to be *uniformly distributed* on the set $S = \{1, 2, 3, \ldots, n\}$ (where $n \geq 1$) if $P(X = k) = 1/n$ for $k = 1, 2, \ldots, n$. What is the mean of X?

7. If $E(X) = 3$, is it true that $E(X^2) = 9$?

8. If $E(X) = 2$, what is $E(E(E(E(X))))$?

9. If $E(X) = 2$ and $E(X^2) = 3$, what is $E(2X^2 - 4X + 1)$?

10. Knowing that $E(X) = 2$ and $E(X^2) = 3$, compute $E(X - X^2 + 7)$.

11. Let X be a random variable with expected value μ. Define $Y = (X - \mu)/\sigma$, where $\sigma \neq 0$. What is the expected value of Y?

12–15 ▪ Given is the probability mass function of a discrete random variable X. Compute $E(X)$, $E(X^2)$, and $E[X(X - 1)]$.

12.

x	$P(X = x)$
-2	0.15
1	0.15
3	0.45
4	0.25

13.

x	$P(X = x)$
0	0.25
1	0.25
2	0.25
3	0.25

14.

x	$P(X = x)$
-2	0.25
-1	0.2
0	0.1
1	0.2
2	0.25

15.

x	$P(X = x)$
0	0.8
1	0.05
2	0.05
3	0.05
4	0.05

16. Find the mean, the median, and the mode of the following data set:

$$20, 16, 20, 27, 30, 20, 18, 29, 30, 30, 27, 22, 18, 30, 20, 16, 14, 32$$
$$24, 18, 27, 22, 20, 20, 20, 20, 18, 18, 30, 27, 20, 30, 25, 30, 28, 24$$

17. Find the mean, the median, and the mode of the following data set:

$$19, 18, 18, 20, 27, 30, 20, 18, 29, 30, 18, 18, 18, 18, 30, 20, 18, 19,$$
$$18, 14, 18, 27, 20, 18, 18, 19, 29, 18, 18, 30, 27, 20, 30, 25, 18, 22$$

18–23 ▪ Consider the following probability mass function of a random variable X. Find each quantity.

x	$P(X = x)$
1	0.2
2	0.4
3	0.3
4	0.1

18. $E(X^2) - (E(X))^2$

19. $E(\sin X) - \sin(E(X))$

20. $E(\ln X)$

21. Geometric mean, i.e., $e^{E(\ln X)}$

22. $E[X(X - E(X))]$

23. $E(1/X)$

24. Assume that the per capita production rate of a population of fish is 1.25 in 70% of years and 0.6 in 30% of years. Calculate the geometric mean and explain what it says about the long-term behaviour of the population.

25. Assume that the per capita production rate of a population of fish is 1.25 in 70% of years and 0.1 in 30% of years. Calculate the geometric mean and explain what it says about the long-term behaviour of the population.

26–29 ▪ Given the histogram of a discrete random variable X, find the mean, the median, and the mode(s).

26.

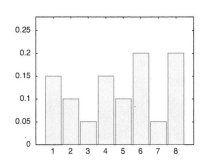

27.

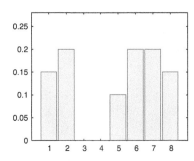

28.

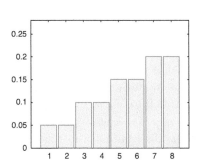

29.

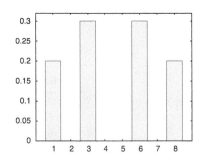

30. Prove the formula $E[E(X)] = E(X)$ directly, without reference to Theorem 7. (Hint: Start by using the formula for the expected value, $E[E(X)] = \sum E(X)P(X = x)$; it's a one-line proof.)

8	The Spread of a Distribution

In Section 7 we learned how to calculate the statistics related to the centre of a distribution. Although important, the mean and the median fail to fully describe a distribution.

For instance, the fact that the mean of a distribution is 5 does not tell us anything about how spread out it is. The mean of both $S_1 = \{3, 4, 5, 6, 7\}$ and $S_2 = \{-20, -10, 5, 20, 30\}$ is 5. However, the largest distance from an element of S_1 to the mean is 2. For S_2, the largest distance is 25.

Both $S_3 = \{0, 1, 6, 1{,}000, 50{,}000\}$ and $S_4 = \{0, 5, 6, 7, 8\}$ have the same median of 6, but differ considerably in the way they are spread out.

We now introduce the most common tools used to describe the **spread** of a distribution: the **range**, the **percentiles**, the **variance**, and the **standard deviation**.

Range, Percentiles, and Boxplots

The *range* of a distribution is the simplest measure of its spread. Usually, we summarize the information about the range by stating the minimum and the maximum values. We say, "The range of S_3 is from 0 to 50,000" or "S_4 ranges from 0 to 8."

We have already met an example of a *percentile*—namely the 50th percentile, or the median. The median splits the distribution into two groups: 50% of the values are smaller than the median, and 50% are larger than the median.

Now we define other important percentiles. As with the median, for a percentile to make sense, we need to have larger distributions with many distinct values.

Definition 24   Percentile

Let X be a random variable. The *pth percentile* (where $0 < p < 1$) is the value of X that is, ideally, larger than p percent of the values.

We comment on why we used the word "ideally" near the end of Example 8.1. The most commonly used percentiles are listed in Table 8.1.

Table 8.1

p	Name and notation
0.25	25th percentile, or first quartile, or lower quartile; Q_1
0.5	50th percentile, or median; M
0.75	75th percentile, or third quartile, or upper quartile; Q_3
0.95	95th percentile

As we just mentioned, the median M divides the values of a random variable into two sets of equal size. The 25th percentile is the median of the set that contains the numbers smaller than M and the 75th percentile is the median of the set of numbers that are larger than M. Thus, the quartiles and the median divide the set of values of a random variable into four roughly equally probable sets.

Example 8.1   **Percentiles for the Milk Production Data**

Recall the daily milk production by 30 Milky Way Farm cows:

$$30, 18, 20, 27, 30, 20, 18, 29, 30, 30, 27, 22, 18, 30, 20,$$
$$24, 18, 27, 22, 29, 29, 20, 29, 18, 18, 30, 27, 20, 30, 25$$

(see Example 7.1). We order the numbers from smallest to largest and list them in two rows, each containing 15 numbers:

$$18, 18, 18, 18, 18, 18, 20, 20, 20, 20, 20, 22, 22, 24, 25,$$
$$27, 27, 27, 27, 29, 29, 29, 29, 30, 30, 30, 30, 30, 30, 30 \qquad (8.1)$$

The minimum is 18, and the maximum is 30. The median is the mean of the last number in the top row and the first number in the bottom row: $(25 + 27)/2 = 26$. The lower quartile (the median of the numbers in the top row) is $Q_1 = 20$, and the upper quartile (the median of the numbers in the bottom row) is $Q_3 = 29$.

To identify the 95th percentile, we note that 95% of 30 (cows) is 28.5. Thus, the 95th percentile is the mean of the amounts of milk produced by the 28th and the 29th cow in the list (8.1), which is 30 L.

Note that the value of 30 is larger than only 23 values; i.e., it's larger than 77%, rather than 95%, of all values. As well, it is larger than or equal to 100%, rather than 95%, of all values. So either way we look at it, strictly speaking, there is no 95th percentile in this distribution. However, since the information could still be (and is) useful, we will refer to 30 L as the 95th percentile. ◢

The comment we made at the end of the previous example applies to all percentiles.

All statistics that we mentioned in Example 8.1 (except for the 95th percentile) form the *five-number summary*.

Definition 25   The Five-Number Summary

The *five-number summary* of a distribution consists of the minimum value of the range, the lower quartile, the median, the upper quartile, and the maximum value of the range. ◢

The five-number summary is often visualized in the form of a *box plot;* see Figure 8.1a. The box plot for the milk production on the Milky Way Farm (Example 8.1) is drawn in Figure 8.1b.

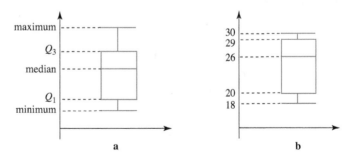

FIGURE 8.1

Box plots

Example 8.2   **Box Plot Diagrams**

Construct the box plot diagrams for the milk production from Example 7.6 in Section 7. For convenience, we repeat the data:

Milkshake Farm:

$$23, 22, 25, 26, 27, 23, 25, 26, 26, 26, 27, 25, 25, 23, 22,$$
$$24, 25, 25, 23, 25, 24, 23, 26, 26, 25, 22, 22, 23, 24, 27$$

Butterscotch Farm:

$$20, 17, 32, 32, 32, 20, 17, 20, 32, 30, 30, 32, 32, 18, 18,$$
$$18, 18, 32, 19, 19, 18, 18, 18, 31, 31, 31, 31, 31, 19, 19$$

▶ Order the Milkshake Farm production numbers:

$$22, 22, 22, 22, 23, 23, 23, 23, 23, 23, 24, 24, 24, 25, 25,$$
$$25, 25, 25, 25, 25, 25, 26, 26, 26, 26, 26, 26, 27, 27, 27$$

The minimum is 22 and the maximum is 27. The median (the mean between the 15th and the 16th numbers in the list) is 25, the lower quartile (the eighth number from either end in the top row) is 23, and the upper quartile (the eighth number from either end in the bottom row) is 26. See Figure 8.2.

In the same way we obtain the statistics for the Butterscotch Farm cows: the minimum is 17 and the maximum is 32. The median is 20, the lower quartile is 18, and the upper quartile is 31; see Figure 8.2. Clearly, the two box plots differ quite a bit.

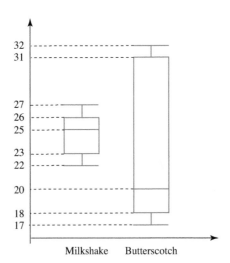

FIGURE 8.2

Box plots of milk production at the two farms

The Variance and the Standard Deviation

To motivate the formulas we will use, we consider two random variables whose distribution functions are given in Tables 8.2 and 8.3.

Table 8.2

x	$P(X = x)$
-1	0.2
0	0.1
1	0.4
2	0.1
3	0.2

Table 8.3

y	$P(Y = y)$
-10	0.2
-7	0.1
1	0.4
9	0.1
12	0.2

Both x and y have the same mean:

$$E(X) = \sum_x x\, P(X = x)$$

$$= (-1)(0.2) + (0)(0.1) + (1)(0.4) + (2)(0.1) + (3)(0.2) = 1$$

and

$$E(Y) = \sum_{y} y\, P(Y = y)$$

$$= (-10)(0.2) + (-7)(0.1) + (1)(0.4) + (9)(0.1) + (12)(0.2) = 1$$

However, Y is more spread out than X. How do we quantify this spread?

One way to do this is to relate the values of the random variable to its mean. To Tables 8.2 and 8.3 we add an extra column, where we calculate the difference between the random variable and its mean. More precisely, we define new random variables $X_1 = X - E(X)$ and $Y_1 = Y - E(Y)$ and display their probability distributions in Tables 8.4 and 8.5.

Table 8.4

x	$P(X = x)$	X_1
-1	0.2	-2
0	0.1	-1
1	0.4	0
2	0.1	1
3	0.2	2

Table 8.5

y	$P(Y = y)$	Y_1
-10	0.2	-11
-7	0.1	-8
1	0.4	0
9	0.1	8
12	0.2	11

What are the means of X_1 and Y_1?

We compute

$$E(X_1) = (-2)(0.2) + (-1)(0.1) + (0)(0.4) + (1)(0.1) + (2)(0.2) = 0$$

Likewise, $E(Y_1) = 0$. What we just discovered is true in general: if X is a random variable and $X_1 = X - E(X)$, then $E(X_1) = 0$ (see Exercise 11).

So, the mean differences $E(X_1)$ and $E(Y_1)$ are equal and cannot help us to distinguish between the spreads of the two distributions. Both are equal to zero, which is not a surprise, since the negative and the positive values cancel each other out.

However, if the differences were all positive, then they would not cancel each other. Given a number, we can make a positive number out of it by squaring, or by taking the absolute value (of course, there are other ways, but the two we mentioned will prove to be most useful). Let's consider squaring.

Consider the squares of the differences, whose distributions are given in Tables 8.6 and 8.7.

Table 8.6

x	$P(X = x)$	X_1^2
-1	0.2	4
0	0.1	1
1	0.4	0
2	0.1	1
3	0.2	4

Table 8.7

y	$P(Y = y)$	Y_1^2
-10	0.2	121
-7	0.1	64
1	0.4	0
9	0.1	64
12	0.2	121

This, time, the means are

$$E(X_1^2) = (4)(0.2) + (1)(0.1) + (0)(0.4) + (1)(0.1) + (4)(0.2) = 1.8 \qquad (8.2)$$

and

$$E(Y_1^2) = (121)(0.2) + (64)(0.1) + (0)(0.4) + (64)(0.1) + (121)(0.2)$$
$$= 61.2 \qquad (8.3)$$

It works! The mean of the differences squared for Y (namely, the mean of Y_1^2) is larger than the corresponding mean for X. This is the quantity we have been looking for, which we now define. Note that the difference squared is the same as the distance squared from the random variable to its mean. Keeping in mind that $X_1^2 = (X - E(X))^2$, the following definition becomes transparent.

Definition 26   The Variance of a Random Variable

Assume that X is a random variable with mean $\mu = E(X)$. The *variance* of X is the real number

$$\text{var}(X) = E\left[(X - \mu)^2\right] = E\left[(X - E(X))^2\right] \qquad \blacktriangle$$

In words, the variance of a random variable X is the mean (expected value) of the distance squared between X and its mean.

If X is a discrete random variable and $p(x) = P(X = x)$ its probability mass function, then

$$\text{var}(X) = \sum_x (x - \mu)^2 P(X = x) = \sum_x (x - \mu)^2 p(x) \qquad (8.4)$$

where the sum is taken over all x for which $p(x) = P(X = x)$ is not zero.

So, in (8.2) and (8.3) we actually calculated the variances: $\text{var}(X) = 1.8$ and $\text{var}(Y) = 61.2$. The larger the variance, the larger is the spread of a distribution.

If $\text{var}(X) = 0$, then (8.4) implies that $(x - \mu)^2 = 0$ for all values of x for which $p(x) > 0$. Thus, the only distribution with $\text{var}(X) = 0$ has one value, $X = \mu$, and $P(X = \mu) = 1$.

Since $\text{var}(X) \geq 0$, it is often denoted by σ^2. Thus, $\sigma^2 = \text{var}(X)$.

Definition 27   Standard Deviation

Let X be a random variable whose variance is $\sigma^2 = \text{var}(X)$. The *standard deviation* of X is the number

$$\sigma = \sqrt{\text{var}(X)} \qquad \blacktriangle$$

Sometimes we use s.d. instead of σ to denote the standard deviation.

If X is measured in some units, then $\text{var}(X)$ is measured in those units squared. Since σ is the square root, the units of σ are the same as the units of X (or its mean μ). That's one of the reasons we introduced the standard deviation.

Example 8.3   Variance and Standard Deviation for the Milk Production

We compare the milk productions on the three farms from the last section, in terms of the spread of the distributions (see Examples 7.1 and 7.8). Recall that the expected daily milk production for all three farms is the same, equal to 24.5.

The distribution for Milky Way Farm is given in Table 8.8.

Table 8.8

x	18	20	22	24	25	27	29	30
$P(X_1 = x)$	6/30	5/30	2/30	1/30	1/30	4/30	4/30	7/30

The variance is

$$\mathrm{var}(X_1) = \sum_x (x - 24.5)^2 P(X_1 = x)$$

$$= (18 - 24.5)^2 \frac{6}{30} + (20 - 24.5)^2 \frac{5}{30} + (22 - 24.5)^2 \frac{2}{30}$$

$$+ (24 - 24.5)^2 \frac{1}{30} + (25 - 24.5)^2 \frac{1}{30} + (27 - 24.5)^2 \frac{4}{30}$$

$$+ (29 - 24.5)^2 \frac{4}{30} + (30 - 24.5)^2 \frac{7}{30}$$

$$= 22.850$$

Thus,

$$\sigma_{X_1} = \sqrt{\mathrm{var}(X_1)} = \sqrt{22.850} \approx 4.780$$

The distribution for the milk production on Milkshake Farm is given in Table 8.9.

Table 8.9

x	22	23	24	25	26	27
$P(X_2 = x)$	4/30	6/30	3/30	8/30	6/30	3/30

As above, we compute $\mathrm{var}(X_2) = 2.450$ and $\sigma_{X_2} = \sqrt{2.450} \approx 1.565$.

From the data for Butterscotch Farm (Table 8.10) we compute $\mathrm{var}(X_3) = 41.849$ and $\sigma_{X_3} = \sqrt{41.849} \approx 6.469$.

Table 8.10

x	17	18	19	20	30	31	32
$P(X_3 = x)$	2/30	7/30	4/30	3/30	2/30	5/30	7/30

In Figure 8.3 we redrew the histograms for the three farms. Clearly, the standard deviation is able to detect the size of the spread: a small standard deviation means a small spread (i.e., the data are clustered around the mean), whereas large values of the standard deviation indicate distributions for which the majority of values are located far from the mean.

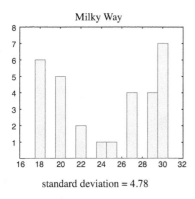
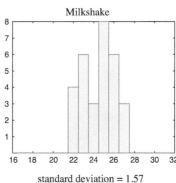
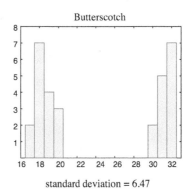

FIGURE 8.3

The standard deviations of the milk production on the three farms

Example 8.4 **Statistics of Surviving Chicks**

Researchers picked a sample of 20 birds that laid the same number of eggs. The following list gives the number of chicks that survived the first month:

$$3, 7, 2, 6, 0, 1, 5, 6, 2, 2, 6, 3, 2, 5, 4, 1, 2, 2, 5, 5$$

Find the mean and the standard deviation of this distribution.

▶ Let X be a random variable that keeps track of how many chicks survived the first month. In Table 8.11 we organize the given list by absolute frequencies and relative frequencies (probabilities).

Table 8.11

x	0	1	2	3	4	5	6	7
Absolute frequencies	1	2	6	2	1	4	3	1
Relative frequencies	1/20	2/20	6/20	2/20	1/20	4/20	3/20	1/20

The expected value of the number of surviving chicks is

$$E(X) = \sum_x x\, P(X = x)$$

$$= 0 \cdot \frac{1}{20} + 1 \cdot \frac{2}{20} + 2 \cdot \frac{6}{20} + 3 \cdot \frac{2}{20} + 4 \cdot \frac{1}{20} + 5 \cdot \frac{4}{20} + 6 \cdot \frac{3}{20} + 7 \cdot \frac{1}{20}$$

$$= \frac{69}{20} = 3.45$$

The variance is

$$\text{var}(X) = \sum_x (x - 3.45)^2 P(X = x)$$

$$= (0 - 3.45)^2 \frac{1}{20} + (1 - 3.45)^2 \frac{2}{20} + (2 - 3.45)^2 \frac{6}{20} + (3 - 3.45)^2 \frac{2}{20}$$

$$+ (4 - 3.45)^2 \frac{1}{20} + (5 - 3.45)^2 \frac{4}{20} + (6 - 3.45)^2 \frac{3}{20} + (7 - 3.45)^2 \frac{1}{20}$$

$$= 3.9475$$

Thus, the standard deviation of this distribution is

$$\sigma_X = \sqrt{\text{var}(X)} = \sqrt{3.9475} \approx 1.9868$$

Next, we explore several properties of the variance.

Theorem 8   Properties of the Variance

Let X be a random variable and a and b be real numbers. Then

(1) $\text{var}(aX) = a^2 \text{var}(X)$

(2) $\text{var}(X + b) = \text{var}(X)$

(3) $\text{var}(X) = E(X^2) - [E(X)]^2$

The first two formulas can be summarized as

$$\text{var}(aX + b) = a^2 \text{var}(X)$$

Formula (2) is intuitively clear: shifting a distribution horizontally by a constant value does not change its spread (think of shifting a histogram). The formula in (3) is an alternative way of calculating the variance and is often more convenient. We'll use it in a moment.

To practise calculations with means and variances, and to review their properties, we prove the theorem.

▶ (1) Using Definition 26, we write

$$\text{var}(aX) = E\left[(aX - E(aX))^2\right]$$

The expected value is linear; thus $E(aX) = aE(X)$, and

$$\begin{aligned}
\text{var}(aX) &= E\left[(aX - aE(X))^2\right] \\
&= E\left[a^2(X - E(X))^2\right] \\
&= a^2 E\left[(X - E(X))^2\right] \\
&= a^2\text{var}(X)
\end{aligned}$$

(We used the linearity of the expected value again when we factored out a^2.)

(2) Using the definition of the variance

$$\text{var}(X + b) = E\left[(X + b - E(X + b))^2\right]$$

(recall that $E(X + b) = E(X) + b$)

$$\begin{aligned}
&= E\left[(X + b - E(X) - b)^2\right] \\
&= E\left[(X - E(X))^2\right] \\
&= \text{var}(X)
\end{aligned}$$

(3) Again, start with Definition 26:

$$\text{var}(X) = E\left[(X - E(X))^2\right]$$

We compute the term in the square brackets

$$(X - E(X))^2 = X^2 - 2E(X)X + (E(X))^2$$

and calculate the expected value of both sides:

$$\begin{aligned}
E\left[(X - E(X))^2\right] &= E\left[X^2 - 2E(X)X + (E(X))^2\right] \\
\text{var}(X) &= E(X^2) - E\left[2E(X)X\right] + E\left[(E(X))^2\right] \\
\text{var}(X) &= E(X^2) - 2E(X) \cdot E(X) + (E(X))^2 \qquad (8.5)
\end{aligned}$$

To simplify the second term on the right side, we applied the formula $E(aX) = aE(X)$ with $a = 2E(X)$. Recall that if b is a real number, then $E(b) = b$. Taking $b = (E(X))^2$, we get $E\left[(E(X))^2\right] = (E(X))^2$, which is how we simplified the third term in (8.5).

Continuing with (8.5), we get

$$\begin{aligned}
\text{var}(X) &= E(X^2) - 2(E(X))^2 + (E(X))^2 \\
&= E(X^2) - (E(X))^2
\end{aligned}$$

Example 8.5   **Alternative Calculation of the Variance**

Use formula (3) from Theorem 8 to recalculate the variance of the random variable X from Example 8.4.

▶ We redraw Table 8.11 by adding the values of X^2; see Table 8.12.

Table 8.12

x	0	1	2	3	4	5	6	7
x^2	0	1	4	9	16	25	36	49
$P(X = x)$	1/20	2/20	6/20	2/20	1/20	4/20	3/20	1/20

In Example 8.4 we calculated the expected value $E(X) = 3.45$. Thus, we only need to find $E(X^2)$.

$$E(X^2) = \sum_x x^2 P(X = x)$$

$$= 0 \cdot \frac{1}{20} + 1 \cdot \frac{2}{20} + 4 \cdot \frac{6}{20} + 9 \cdot \frac{2}{20} + 16 \cdot \frac{1}{20} + 25 \cdot \frac{4}{20} + 36 \cdot \frac{3}{20} + 49 \cdot \frac{1}{20}$$

$$= \frac{317}{20} = 15.85$$

It follows that

$$\text{var}(X) = E(X^2) - (E(X))^2 = 15.85 - 3.45^2 = 3.9475$$

Remarks

(1) Note that $E(X^2) \neq (E(X))^2$. From $\text{var}(X) = E(X^2) - (E(X))^2 \geq 0$ we conclude that $E(X^2) \geq (E(X))^2$. The equality holds only when $\text{var}(X) = 0$.

(2) Instead of squaring the differences between the values of a random variable and its mean, we can take the absolute value. In that case, we obtain the *mean absolute deviation* (MAD)

$$\text{MAD} = E\left(|X - E(X)|\right)$$

Although MAD works (i.e., distinguishes distributions based on the spread; see Exercise 25), it is not as commonly used as the variance. We will not use it in this book.

(3) Assume that a random variable X takes on the values $x_1, x_2, \ldots, x_n$ with equal probability. Thus, $P(X = x_i) = 1/n$ for all $i = 1, 2, \ldots, n$. The expected value

$$E(X) = \sum_{i=1}^{n} x_i P(X = x_i) = \sum_{i=1}^{n} x_i \frac{1}{n} = \frac{x_1 + x_2 + \cdots + x_n}{n}$$

corresponds to our usual notion of what the mean ("average value") is. The variance is

$$\text{var}(X) = \sum_{i=1}^{n} (x_i - E(X))^2 P(X = x_i) = \frac{1}{n} \sum_{i=1}^{n} (x_i - E(X))^2$$

and the standard deviation is

$$\sigma = \sqrt{\frac{1}{n} \sum_{i=1}^{n} (x_i - E(X))^2} \tag{8.6}$$

This formula differs from the standard deviation formula

$$\sigma = \sqrt{\frac{1}{n-1} \sum_{i=1}^{n} (x_i - E(X))^2} \tag{8.7}$$

that is implemented in many mathematics and statistics software packages and programmable calculators. We will not go into reasons why there are two different formulas (let's just say that it's related to the fact that we are calculating the standard deviation of a *sample* of the whole population and not of the whole population; this issue is addressed in statistics courses). We will not use formula (8.7) in this book.

If we decide to use mathematics or statistics software, we need to check which formula is actually being used. If (8.7) is used, then we have to multiply the answer by $\sqrt{(n-1)/n}$ to obtain the answer that corresponds to (8.6).

Summary

Several statistics are used to measure the **spread** of a distribution. The **pth percentile** is the value of the random variable that is larger than p percent of all values of X. The **lower quartile** (25th percentile), the **median** (50th percentile), and the **upper quartile** (75th percentile), together with the **minimum** and the **maximum** values of the range of X, form the **five-number summary** statistics of X. Often, the five-number summary is represented in a **box plot diagram.** The **variance** is the expected value of the distance squared from the mean, and the **standard deviation** is the square root of the variance. The variance and the standard deviation are the most common statistics used to quantify the spread of a distribution.

8	Exercises

1. Order the following three samples based on their standard deviations (from the smallest to the largest): $A = \{2, 2, 3, 4, 4\}$, $B = \{2, 3, 3, 3, 4\}$, $C = \{1, 1, 3, 5, 5\}$.

2. Order the following three samples based on their standard deviations (from the smallest to the largest): $A = \{-2, -1, 0, 1, 2\}$, $B = \{-3, 0, 0, 0, 3\}$, $C = \{-4, 0, 1, 1, 2\}$.

3. It turns out that the variance of a random variable X is 22, but you want it to be 2. How would you change the values of X to make it so?

4. Suppose that $\text{var}(X) = 5$. Using X, define a random variable Y such that $\text{var}(Y) = 10$.

5. A random variable X is said to be *uniformly distributed* on the set $S = \{-4, -3, -2, -1, 0, 1, 2, 3, 4\}$ if $P(X = k) = 1/9$ for $k = 1, 2, \ldots, 9$. What is the variance of X?

6. A random variable X is said to be *uniformly distributed* on the set $S = \{-3, -2, -1, 0, 1, 2, 3\}$ if $P(X = k) = 1/7$ for $k = 1, 2, \ldots, 7$. What is the variance of X?

▷ 7–10 ▪ Find the variance and the standard deviation of each distribution.

7.

x	$P(X = x)$
0	0.15
1	0.15
2	0.15
4	0.55

8.

x	$P(X = x)$
0	0.1
1	0.4
2	0.4
3	0.1

9.

x	$P(X = x)$
−2	0.25
−1	0.2
0	0.1
1	0.2
2	0.25

10.

x	$P(X = x)$
0	0.8
1	0.05
2	0.05
3	0.05
4	0.05

11. Show that if $X_1 = X - E(X)$, then $E(X_1) = 0$. (Hint: Try to prove it directly, or use Theorem 7 in Section 7; if you do, it's a one-line proof.)

12. Assume that X is a random variable with standard deviation σ. Define $Y = (X - a)/\sigma$, where a is a real number. What is the standard deviation of Y?

13. Using $\text{var}(X) = E(X^2) - [E(X)]^2$ and the properties of the expected value, prove that $\text{var}(aX) = a^2\text{var}(X)$ for a real number a.

14. Using $\text{var}(X) = E(X^2) - [E(X)]^2$ and the properties of the expected value, prove that $\text{var}(X + b) = \text{var}(X)$ for a real number b.

15. Draw the box plot diagrams for the systolic blood pressure measurements (mmHg) for the following two samples and compare them: sample of 12 healthy adults: 120, 123, 138, 110, 125, 128, 140, 132, 138, 116, 122, 125; sample of 12 adults with a history of cardiovascular problems: 136, 142, 150, 148, 160, 154, 162, 166, 154, 154, 160, 158.

16. Draw the box plot diagrams for the blood glucose level measurements (mmol/L) for the following two samples and compare them: sample of 12 healthy adults: 4.2, 3.6, 3.5, 4, 4.1, 4.2, 4.2, 4, 3.8, 3.8, 3.8, 3.4; sample of 12 adults who have experienced problems related to increased blood glucose levels: 4, 4.4, 6, 5.8, 6, 4.4, 5, 6.2, 6.8, 6, 6, 6.2.

▽ 17–20 ▪ For each data set, calculate the five-number summary and draw a box plot.

17. The lifespan of lions in captivity (in years): 18, 17, 19, 24, 22, 20, 20, 16, 14, 24, 22, 25, 24, 20.

18. The lifespan of lions in the wild (in years): 10, 8, 12, 14, 18, 12, 16, 14, 14, 12, 11, 15.

19. The lifespan of moose in Algonquin Provincial Park (in years): 12, 20, 24, 26, 25, 23, 21, 20, 28, 20, 24, 23, 27, 25.

◣ 20. The lifespan of mosquitoes (in weeks): 4.5, 5, 6, 6.5, 7, 8, 5, 5.5, 6, 6, 8, 7.5, 6.

▽ 21–24 ▪ Given the histogram of a discrete random variable, find its variance and standard deviation.

21.

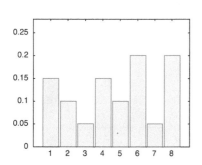

22.

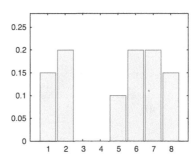

23.

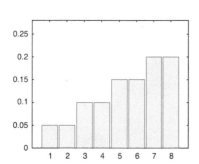

24.

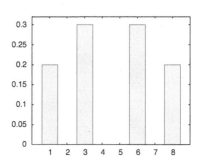

◣

25. Using the milk production data in Tables 8.8 to 8.10 in Example 8.3, calculate the mean absolute deviations (MAD) for the three populations. Convince yourself that the MAD is able to detect the differences in their spreads.

26. Consider the data set of the lifespan (in years) of lions in captivity

$$C = \{18, 17, 19, 24, 22, 20, 20, 16, 14, 24, 22, 25, 24, 20\}$$

and the data set of the lifespan (in years) of lions in the wild

$$W = \{10, 8, 12, 14, 18, 12, 16, 14, 14, 12, 11, 15\}$$

Compute the standard deviations and the mean absolute deviations of C and W. Do the two statistics agree on detecting a larger/smaller spread?

9	Joint Distributions

So far we have studied one random variable and its associated distribution. In order to study two variables and to investigate their relationship, we introduce the concept of a **joint distribution.** The joint distribution gives the probability of each pair of values of the random variables involved.

From the joint distribution we extract marginal distributions that describe each random variable separately. As well, we define the conditional distribution, i.e., the probability of one random variable given that the other variable takes on a particular value.

Joint Distribution

We consider an example first.

Example 9.1   Joint Distribution of Populations of Squirrels

Two species of squirrels, grey and brown, live in a forest. Researchers measured the lengths of their tails and classified them as short, medium, or long. Define the random variables

$$X = \begin{cases} 1 & \text{squirrel is brown} \\ 2 & \text{squirrel is grey} \end{cases}$$

and

$$Y = \begin{cases} 1 & \text{squirrel has a short tail} \\ 2 & \text{squirrel has a medium-length tail} \\ 3 & \text{squirrel has a long tail} \end{cases}$$

The data obtained from 1,000 squirrels are summarized in Table 9.1.

Table 9.1

	$X = 1$	$X = 2$	Total
$Y = 1$	150	90	240
$Y = 2$	220	340	560
$Y = 3$	50	150	200
Total	420	580	1,000

For instance, 220 brown squirrels have medium-length tails ($X = 1$ and $Y = 2$) and 150 grey squirrels have long tails ($X = 2$ and $Y = 3$). Replacing the frequencies in Table 9.1 with relative frequences, we obtain Table 9.2:

Table 9.2

	$X = 1$	$X = 2$	Total
$Y = 1$	0.15	0.09	0.24
$Y = 2$	0.22	0.34	0.56
$Y = 3$	0.05	0.15	0.2
Total	0.42	0.58	1

Reading from Table 9.2: the probability that a randomly chosen squirrel is brown and has a medium-length tail is $P(X = 1, Y = 2) = 0.22$.

Likewise, $P(X = 2, Y = 3) = 0.15$ is the probability that a randomly chosen squirrel is grey and has a long tail.

The probability that a randomly chosen squirrel has a medium-length tail is 0.56, and the probability that a randomly chosen squirrel is brown is 0.42.

The distribution given in Table 9.2 is called the *joint probability distribution*. It assigns a probability to every combination of the values for X and Y. We write

$$p(x, y) = P(X = x, Y = y)$$

Note that the horizontal and the vertical totals in Table 9.2 are 1. The vertical sum $(0.24 + 0.56 + 0.2 = 1)$ represents the fact that the probability that a squirrel has a short, or a medium-length, or a long tail is 1. Likewise, $0.42 + 0.58 = 1$ because a squirrel must be either brown or grey.

Definition 28   Joint Probability Distribution

Assume that X and Y are discrete random variables with ranges $x_1, x_2, \ldots, x_m$ and $y_1, y_2, \ldots, y_n$, respectively. The probabilities

$$P(x_i, y_j) = P(X = x_i, Y = y_j)$$

$(i = 1, 2, \ldots, m$ and $j = 1, 2, \ldots, n)$ of each composite event $X = x_i$ and $Y = y_j$ define the *joint probability distribution* of X and Y.

As in the example preceding the definition, we usually describe a joint probability distribution in a table; see Table 9.3.

Table 9.3

	$X = x_1$	$\cdots$	$X = x_i$	$\cdots$	$X = x_m$
$Y = y_1$	$P(X = x_1, Y = y_1)$		$P(X = x_i, Y = y_1)$		$P(X = x_m, Y = y_1)$
$\cdots$	$\cdots$		$\cdots$		$\cdots$
$Y = y_j$	$P(X = x_1, Y = y_j)$		$P(X = x_i, Y = y_j)$		$P(X = x_m, Y = y_j)$
$\cdots$	$\cdots$		$\cdots$		$\cdots$
$Y = y_n$	$P(X = x_1, Y = y_n)$		$P(X = x_i, Y = y_n)$		$P(X = x_m, Y = y_n)$

By extending Definition 28 we can define a joint probability distribution for any number of random variables. For instance, the joint probability distribution of three random variables X, Y, and Z consists of the probabilities

$$P(X = x, Y = y, Z = z)$$

where x, y, and z are all possible values of X, Y, and Z, respectively.

We look a bit more closely at the totals in Table 9.2. The value $Y = 1$ represents squirrels with short tails. We compute

$$P(Y = 1) = P((Y = 1 \text{ and } X = 1) \text{ or } (Y = 1 \text{ and } X = 2))$$
$$= P(Y = 1 \text{ and } X = 1) + P(Y = 1 \text{ and } X = 2)$$
$$= 0.15 + 0.09 = 0.24$$

The events "$Y = 1$ and $X = 1$" (brown squirrel, short tail) and "$Y = 1$ and $X = 2$" (grey squirrel, short tail) are mutually exclusive. That's why we calculated the probability above as the sum of the probabilities. As well, the probability that a randomly chosen squirrel has a medium-length tail is

$$P(Y = 2) = P((Y = 2 \text{ and } X = 1) \text{ or } (Y = 2 \text{ and } X = 2))$$
$$= P(Y = 2 \text{ and } X = 1) + P(Y = 2 \text{ and } X = 2)$$
$$= 0.22 + 0.34 = 0.56$$

The probability that a randomly chosen squirrel is brown is
$$P(X = 1) = P((X = 1 \text{ and } Y = 1) \text{ or } (X = 1 \text{ and } Y = 2)$$
$$\text{or } (X = 1 \text{ and } Y = 3))$$
$$= P(X = 1 \text{ and } Y = 1) + P(X = 1 \text{ and } Y = 2)$$
$$+ P(X = 1 \text{ and } Y = 3)$$
$$= 0.15 + 0.22 + 0.05 = 0.42$$

Thus, the rows and the columns in Table 9.2 represent individual random variables. With this in mind, we redraw Table 9.2 by identifying the totals in the last column and in the last row by the probabilities they represent; see Table 9.4.

Table 9.4

	$X = 1$	$X = 2$	Total
$Y = 1$	0.15	0.09	$P(Y = 1) = 0.24$
$Y = 2$	0.22	0.34	$P(Y = 2) = 0.56$
$Y = 3$	0.05	0.15	$P(Y = 3) = 0.2$
Total	$P(X = 1) = 0.42$	$P(X = 2) = 0.58$	1

By adding up all of the entries in the ith column in Table 9.3, we get
$$P(X = x_i) = P((X = x_i \text{ and } Y = y_1) \text{ or } (X = x_i \text{ and } Y = y_2)$$
$$\text{or } \ldots \text{ or } (X = x_i \text{ and } Y = y_n))$$
$$= P(X = x_i \text{ and } Y = y_1) + P(X = x_i \text{ and } Y = y_2)$$
$$+ \cdots + P(X = x_i \text{ and } Y = y_n)$$

The probability that $Y = y_j$ is the sum of the terms in the jth row:
$$P(Y = y_j) = P(X = x_1 \text{ and } Y = y_j) + P(X = x_2 \text{ and } Y = y_j)$$
$$+ \cdots + P(X = x_m \text{ and } Y = y_j)$$

In this way, using the joint probability distribution, we are able to find the probability distributions of both X and Y.

Definition 29 Marginal Probability Distribution

Assume that $p(x, y) = P(X = x, Y = y)$ represents the joint probability distribution of random variables X and Y, and that the range of X is $x_1, x_2, \ldots, x_m$ and the range of Y is $y_1, y_2, \ldots, y_n$ $(m, n \geq 1)$.

The *marginal probability distribution* p_X of X is given by
$$p_X(x_i) = P(X = x_i) = \sum_{j=1}^{n} P(X = X_i, Y = y_j)$$

for $i = 1, 2, \ldots, m$. The *marginal probability distribution* p_Y of Y is given by
$$p_Y(y_j) = P(Y = y_j) = \sum_{i=1}^{m} P(X = X_i, Y = y_j)$$

where $i = 1, 2, \ldots, m$.

Example 9.2 Joint and Marginal Distributions

Consider the joint probability distribution $p(x, y)$ given in Table 9.5. Find $p(1, -1)$ and $p(0, 2)$. Identify the marginal probability distributions $p_X(x)$ and $p_Y(y)$.

Table 9.5

	$X = -1$	$X = 0$	$X = 1$
$Y = -2$	0.15	0.05	0.2
$Y = -1$	0.08	0.15	0.2
$Y = 2$	0.07	0.05	0.05

▶ By definition,

$$p(1, -1) = P(X = 1, Y = -1) = 0.2$$

and

$$p(0, 2) = P(X = 0, Y = 2) = 0.05$$

Since the range of X is $\{-1, 0, 1\}$, in order to describe the marginal probability distribution $p_X(x)$ we need to calculate $P(X = -1)$, $P(X = 0)$, and $P(X = 1)$. Using Definition 29,

$$P(X = -1) = P(X = -1, Y = -2) + P(X = -1, Y = -1) + P(X = -1, Y = 2)$$
$$= 0.15 + 0.08 + 0.07 = 0.3$$

As well,

$$P(X = 0) = P(X = 0, Y = -2) + P(X = 0, Y = -1) + P(X = 0, Y = 2)$$
$$= 0.05 + 0.15 + 0.05 = 0.25$$

In the same way (by adding up the entries in the last column) we get $P(X = 1) = 0.45$. The marginal probability distribution for X is given in Table 9.6.

By adding up the probabilities along the rows of Table 9.5 we obtain the marginal probability distribution $p_Y(y)$; see Table 9.7. ◢

Table 9.6

x	$P(X = x)$
-1	0.3
0	0.25
1	0.45

Table 9.7

y	$P(Y = y)$
-2	0.4
-1	0.43
2	0.17

Recall that the joint probability distribution is given by

$$p(x, y) = P(X = x \text{ and } Y = y)$$

(To make the notation cleaner, we drop subscripts; so x represents all x_i, $i = 1, 2, \ldots, m$, and y represents all y_j, $j = 1, 2, \ldots, n$.)

Assuming that $X = x$ and $Y = y$ are independent events, we get

$$p(x, y) = P(X = x \text{ and } Y = y) = P(X = x)P(Y = y)$$

If this is true for all x in the range of X and all y in the range of Y, then X and Y are called *independent* random variables.

Reading the following definition, keep in mind that $P(X = x)$ is the marginal distribution $p_X(x)$ and $P(Y = y)$ is the marginal distribution $p_Y(y)$.

Definition 30   Independent Random Variables

Assume that X and Y are discrete random variables. If the joint probability distribution is equal to the product of the marginal distributions of X and Y, then X and Y are called *independent* random variables. ◢

Thus, X and Y are independent if

$$p(x, y) = p_X(x)p_Y(y)$$

for all x in the range of X and for all y in the range of Y.

Example 9.3 Independent Random Variables

Consider two random variables X and Y whose joint probability distribution is given in Table 9.8. Are X and Y independent?

Table 9.8

	$X = 1$	$X = 2$
$Y = -1$	0.04	0.16
$Y = 1$	0.16	0.64

▶ We need to check whether

$$P(X = x, Y = y) = P(X = x)P(Y = y) \tag{9.1}$$

is true for all combinations of $x = 1$ or $x = 2$ and $y = -1$ or $y = 1$.

To check the right side in (9.1) we need to know all of the marginal probabilities. Therefore, we append a row and a column to Table 9.8 and calculate the marginal probabilities by adding up entries in each row and in each column; see Table 9.9.

Table 9.9

	$X = 1$	$X = 2$	
$Y = -1$	0.04	0.16	$P(Y = -1) = 0.2$
$Y = 1$	0.16	0.64	$P(Y = 1) = 0.8$
	$P(X = 1) = 0.2$	$P(X = 2) = 0.8$	

We see that $P(X = 1, Y = -1) = 0.04$ is equal to

$$P(X = 1)P(Y = -1) = (0.2)(0.2) = 0.04$$

As well, $P(X = 1, Y = 1) = 0.16$ and

$$P(X = 1)P(Y = 1) = (0.2)(0.8) = 0.16$$

In the same way we check that $P(X = 2, Y = -1) = 0.16$ is equal to

$$P(X = 2)P(Y = -1) = (0.8)(0.2) = 0.16$$

and $P(X = 2, Y = 1) = 0.64$ is equal to

$$P(X = 2)P(Y = 1) = (0.8)(0.8) = 0.64$$

We showed that (9.1) holds for all possible combinations of the values of X and the values of Y. Consequently, X and Y are independent. ◣

Example 9.4 Non-independent Random Variables

Show that the random variables X and Y in Example 9.1 are not independent.

▶ To prove that X and Y are independent, we need to verify that

$$P(X = x, Y = y) = P(X = x)P(Y = y) \tag{9.2}$$

holds for *all* combinations of the values of X and the values of Y. However, to prove that X and Y are not independent, we need to show that (9.2) fails to hold for *one* value of X and *one* value of Y.

All we need is Table 9.4. We read that $P(X = 1, Y = 2) = 0.22$. Since

$$P(X = 1)P(Y = 2) = (0.42)(0.56) = 0.2352$$

is not equal to 0.22, it follows that X and Y are not independent. ◣

Example 9.5   Calculations with Joint and Marginal Distributions

Table 9.10 shows the probability mass functions for random variables X and Y. Assuming that X and Y are independent, find

(a) The probability that $X = 3$ and $Y = 4$.

(b) The probability that X is even and $Y = 3$.

(c) The probability that $X = Y$.

Table 9.10

x, y	$P(X = x)$	$P(Y = y)$
1	0.4	0.6
2	0.15	0.05
3	0.4	0.05
4	0.05	0.3

▶ (a) The assumption on independence implies that

$$P(X = 3, Y = 4) = P(X = 3)P(Y = 4) = (0.4)(0.3) = 0.12$$

(b) We are asked to find the probability of the event $(X = 2$ or $X = 4)$ and $Y = 3$. By mutual exclusivity, and then by independence, we get

$$\begin{aligned}
P((X = 2 &\text{ or } X = 4) \text{ and } Y = 3) \\
&= P((X = 2 \text{ and } Y = 3) \text{ or } (X = 4 \text{ and } Y = 3)) \\
&= P(X = 2 \text{ and } Y = 3) + P(X = 4 \text{ and } Y = 3) \\
&= P(X = 2)P(Y = 3) + P(X = 4)P(Y = 3) \\
&= (0.15)(0.05) + (0.05)(0.05) = 0.01
\end{aligned}$$

(c) The event $X = Y$ can be written as $\{(X = 1$ and $Y = 1)$ or $(X = 2$ and $Y = 2)$ or $(X = 3$ and $Y = 3)$ or $(X = 4$ and $Y = 4)\}$. Thus

$$\begin{aligned}
P(X = Y) &= P(X = 1 \text{ and } Y = 1) + P(X = 2 \text{ and } Y = 2) \\
&\quad + P(X = 3 \text{ and } Y = 3) + P(X = 4 \text{ and } Y = 4) \\
&= P(X = 1)P(Y = 1) + P(X = 2)P(Y = 2) + P(X = 3)P(Y = 3) + \\
&\quad + P(X = 4)P(Y = 4) \\
&= (0.4)(0.6) + (0.15)(0.05) + (0.4)(0.05) + (0.05)(0.3) \\
&= 0.2825
\end{aligned}$$

Given two random variables X and Y, we define the *conditional probability* in the same way as in Section 4:

$$P(X = x \,|\, Y = y) = \frac{P(X = x \text{ and } Y = y)}{P(Y = y)} = \frac{P(X = x, Y = y)}{P(Y = y)}$$

Example 9.6   Conditional Probability with Squirrels

Going back to Example 9.1 and the probabilities in Table 9.4, find

(a) The probability that a randomly chosen brown squirrel has a long tail.

(b) The probability that a randomly chosen medium-tailed squirrel is grey.

▶ (a) We are asked to find

$$P(\text{long tail} \mid \text{brown squirrel}) = P(Y = 3 \mid X = 1)$$
$$= \frac{P(X = 1, Y = 3)}{P(X = 1)}$$
$$= \frac{0.05}{0.42} = \frac{5}{42} \approx 0.119$$

(b) We need to find

$$P(\text{grey} \mid \text{medium-tailed squirrel}) = P(X = 2 \mid Y = 2)$$
$$= \frac{P(X = 2, Y = 2)}{P(Y = 2)}$$
$$= \frac{0.34}{0.56} \approx 0.607$$

Two important formulas related to independent random variables are given in the following theorem.

Theorem 9   Properties of Independent Random Variables

Assume that X and Y are independent discrete random variables. Then

(1) $E(XY) = E(X)E(Y)$

(2) $\text{var}(X + Y) = \text{var}(X) + \text{var}(Y)$.

The proof of (1) is discussed in Exercise 35. To prove (2), we start with formula (3) from Theorem 8 in Section 8:

$$\begin{aligned}
\text{var}(X + Y) &= E\left[(X + Y)^2\right] - [E(X + Y)]^2 \\
&= E(X^2 + 2XY + Y^2) - [E(X) + E(Y)]^2 \\
&= E(X^2) + 2E(XY) + E(Y^2) - (E(X))^2 - 2E(X)E(Y) - (E(Y))^2 \\
&= E(X^2) - (E(X))^2 + E(Y^2) - (E(Y))^2 \\
&= \text{var}(X) + \text{var}(Y)
\end{aligned}$$

(Note that by (1), $2E(XY) - 2E(X)E(Y) = 0$.)

We will use these formulas (especially formula (2)) in the forthcoming sections.

Summary   In order to describe two (or more) random variables simultaneously, we build a **joint distribution.** In the case of two random variables, the joint distribution gives the probability of the occurrence of each pair of values of the two variables. Adding up all probabilities of one random variable, we obtain the **marginal probability distribution** of the other. If the joint probability distribution is equal to the product of the marginal distributions, then the random variables are **independent.** We define the **conditional probability** in the same way as in Section 4.

9	Exercises

1. Assume that X and Y are independent random variables with distributions $P(X = 1) = 0.2$, $P(X = 2) = 0.8$ and $P(Y = 1) = 0.7$, $P(Y = 2) = 0.3$. Find the joint probability distribution of X and Y.

2. Assume that X and Y are independent random variables with distributions $P(X = 1) = 0.5$, $P(X = 2) = 0.5$ and $P(Y = 1) = 0.5$, $P(Y = 2) = 0.5$. Find the joint probability distribution of X and Y.

3–4 ▪ Fill in the missing entries in the joint probability distribution so that X and Y become independent random variables.

3.

	$X = 0$	$X = 1$
$Y = 0$	0.1	0.3
$Y = 1$		

4.

	$X = 0$	$X = 1$
$Y = 0$		0.25
$Y = 1$		0.35

5. Assume that X and Y are independent random variables with distributions $P(X = 1) = 0.2$, $P(X = 2) = 0.8$ and $P(Y = 1) = 0.9$, $P(Y = 2) = 0.1$. Find the joint probability distribution of X and Y. Calculate $P(X = 1 \mid Y = 1)$, $P(X = 1 \mid Y = 2)$, and $P(X = 1)$ and explain the relationship between the three probabilities.

6. Assume that X and Y are independent random variables with distributions $P(X = 1) = 0.45$, $P(X = 2) = 0.55$ and $P(Y = 1) = 0.55$, $P(Y = 2) = 0.45$. Find the joint probability distribution of X and Y. Calculate $P(X = 2 \mid Y = 1)$, $P(X = 2 \mid Y = 2)$, and $P(X = 2)$ and explain the relationship between the three probabilities.

7–10 ▪ Consider the joint distribution of blood types in Canada, where the random variables are $G =$ "blood group" and $R =$ "Rhesus factor." (Source: Canadian Blood Services.)

	$G = A$	$G = B$	$G = AB$	$G = O$
$R = +$	0.36	0.076	0.025	0.39
$R = -$	0.06	0.014	0.005	0.07

7. If someone's blood group is B, what is the probability that their blood type is B+?

8. What is the probability that a person whose Rhesus factor is negative has blood group AB?

9. Find the distribution of R conditional on $G = B$.

10. Find the distribution of G conditional on $R = +$.

11–14 ▪ Consider the joint distribution related to testing for an allergy to bee venom. The random variables are $T =$ "test result" and $A =$ "person has an allergy to bee venom."

	$T =$ positive	$T =$ negative	$T =$ inconclusive
$A =$ allergy	0.3	0.07	0.1
$A =$ no allergy	0.03	0.45	0.05

11. Find the two marginal distributions and explain what they mean.

12. What is the probability that someone who tests positive has the allergy?

13. What is the probability that someone who tests negative is actually allergic to bee venom?

14. What is the probability that someone who is allergic to bee venom will have an inconclusive test?

15. Consider two random variables: X, with range $\{1, 2\}$, and Y, with range $\{3, 4\}$. Find the joint distribution of the two random variables knowing that $P(X = 1) = 0.4$, $P(X = 2) = 0.6$, $P(Y = 3 \mid X = 1) = 0.7$, and $P(Y = 3 \mid X = 2) = 0.1$.

16. Consider two random variables: X, with range $\{1, 2\}$, and Y, with range $\{3, 4\}$. Find the joint distribution of the two random variables knowing that $P(Y = 3) = 0.1$, $P(Y = 4) = 0.9$, $P(X = 1 \mid Y = 3) = 0.2$, and $P(X = 1 \mid Y = 4) = 0.4$.

17–22 ▪ The joint distribution below describes the feeding patterns of three animal species. The random variables are $P = $ "predator" and $F = $ "food."

	$P = $ Brown bear	$P = $ Wolf	$P = $ Fox
$F = $ Fish	0.2	0.02	0.03
$F = $ Insects	0.1	0.05	0.05
$F = $ Small mammals	0.2	0.25	0.1

17. Find the two marginal distributions.

18. The sum of the numbers in the first row is 0.25. Dividing the entries in the first row by 0.25, we get 0.2/0.25, 0.02/0.25, and 0.03/0.25. What conditional probabilities do these three numbers represent? What is their sum? Why?

19. The sum of the numbers in the second column is 0.32. Dividing the entries in the second column by 0.32, we get 0.02/0.32, 0.05/0.32, and 0.25/0.32. What conditional probabilities do these three numbers represent? What is their sum? Why?

20. What is the probability that a wolf will prey on a fish?

21. What is the probability that a bear will prey on a small mammal?

22. A small mammal has been caught. What is the probability that it was caught by a wolf?

23–26 ▪ For each joint distribution:

(a) Find both marginal distributions.

(b) Determine whether or not X and Y are independent.

23.

	$X = 0$	$X = 1$
$Y = 0$	0.05	0.1
$Y = 1$	0.45	0.4

24.

	$X = 0$	$X = 1$
$Y = 0$	0.09	0.21
$Y = 1$	0.21	0.49

25.

	$X = 0$	$X = 1$
$Y = 0$	0.12	0.18
$Y = 1$	0.22	0.28
$Y = 2$	0.02	0.18

26.

	$X = 0$	$X = 1$
$Y = 0$	0.2	0.3
$Y = 1$	0.08	0.12
$Y = 2$	0.12	0.18

27. Consider the joint distribution given in Exercise 26. Find $P(Y = 0 \mid X = 0)$ and $P(Y = 0 \mid X = 1)$. What is the sum $P(Y = 0 \mid X = 0) + P(Y = 0 \mid X = 1)$? Why?

28. Consider the joint distribution given in Exercise 26. What is the sum $P(X = 0 \mid Y = 0) + P(X = 0 \mid Y = 1) + P(X = 0 \mid Y = 2)$? Why?

29–32 ▪ Consider the following joint distribution.

	$X = 0$	$X = 1$
$Y = 0$	0.05	0.1
$Y = 1$	0.1	0.1
$Y = 2$	0.4	0.25

29. Find the marginal probability distribution of X.

30. Find the marginal probability distribution of Y.

31. Find the distribution of X conditional on $Y = 2$.

32. Find the distribution of Y conditional on $X = 0$.

33–34 ▪ Consider the following joint distribution.

	$Y = 1$	$Y = 2$
$X = -2$	0	0.12
$X = -1$	0.1	0.38
$X = 0$	0.26	0.14

33. Find the marginal probability distributions of X and Y.

34. Find the distribution of X conditional on $Y = 1$. Find the distribution of Y conditional on $X = -2$.

35. Assume that X is a discrete random variable with range $\{1, 2\}$ and Y is a discrete random variable with range $\{3, 4, 5\}$. Assuming that X and Y are independent, show that $E(XY) = E(X)E(Y)$. Generalize your proof: assuming that X and Y are independent, and that the range of X is $\{x_1, x_2, \ldots, x_m\}$ and the range of Y is $\{y_1, y_2, \ldots, y_n\}$, prove that $E(XY) = E(X)E(Y)$.

10 The Binomial Distribution

Having learned about distributions and how to assign probabilities, we now turn our attention to one special distribution, the **binomial distribution,** that arises in numerous models in biology and elsewhere. The binomial distribution is based on the concept of a Bernoulli experiment, which is the opening topic of this section.

Bernoulli Experiment and Bernoulli Random Variable

We define a simple random variable with two outcomes.

Definition 31   **Bernoulli Random Variable**

A discrete random variable that takes on the value 1 with probability p and the value 0 with probability $1 - p$ is called a *Bernoulli random variable.*   ◣

Table 10.1

b	$P(B = b)$
0	$1 - p$
1	p

A random experiment whose outcomes can be described using a Bernoulli random variable is called a *Bernoulli experiment* or a *Bernoulli trial.* The value 1 is often called a "success," and p is the probability of success.

The meaning of 0 and 1 (or success and no-success) depends on the particular experiment. The probability distribution of the Bernoulli random variable B is shown in Table 10.1.

Example 10.1   Bernoulli Experiment: Tossing a Coin

Define the random variable B to count the number of tails in a single flip of a coin. We can think of B as a Bernoulli random variable, where $B = 1$ represents flipping tails and $B = 0$ represents flipping heads. So, flipping a coin once is a Bernoulli experiment for which success is flipping tails. In the case of a fair coin, $p = P(B = 1) = 0.5$ and $P(B = 0) = 1 - p = 0.5$.   ◣

Example 10.2   Bernoulli Experiment: Occurrence of a Virus

Let V track the occurrence of a virus during a given time interval (say, one month) within a certain population. The outcome $V = 1$ indicates the presence of the virus (so, in this context, that's success), and $V = 0$ indicates that there is no virus. Assume that $p = P(V = 1) = 0.2$ is the probability that the virus is present. The virus is absent with probability $P(V = 0) = 1 - p = 0.8$. In this way, V becomes a Bernoulli random variable.   ◣

Example 10.3   Bernoulli Experiment: Finding Artifacts

The probability that, on a given day, a team of archaeologists finds an important (scientifically valuable) artifact is 0.015. Define

$$B = \begin{cases} 1 & \text{important artifact is found (success)} \\ 0 & \text{important artifact is not found} \end{cases}$$

This is an example of a Bernoulli trial (experiment), and B is a Bernoulli random variable.

Table 10.2

b	$P(B = b)$
0	0.985
1	0.015

The probability distribution of B is given in Table 10.2.   ◣

Take a Bernoulli experiment and keep repeating it. Assume that the repetitions are independent of each other, i.e., the outcome of one experiment does not affect the outcomes of any other experiment. There are all kinds of questions that we would like to know the answer to.

Example 10.4   Questions That We Would Like to Know the Answers To

(i) Assume that we repeat the experiment in Example 10.1 ten times (i.e., we toss a coin ten times). What is the probability that we flip exactly seven tails?

(ii) Related to Example 10.2, what is the probability that there are exactly 8 virus-free months in a year?

(iii) What is the probability that the archeologists in Example 10.3 find important artifacts more often than 1 day in a month?

We can answer these questions by understanding how a *binomial distribution* works.

The Binomial Distribution

Assume that we repeat the same Bernoulli experiment and that the outcomes are independent. Denote by N the random variable that counts the number of successes (i.e., the number of outcomes $B = 1$) in n repetitions of the experiment. Since the range of N is the set $\{0, 1, 2, 3, \dots, n\}$, we need to figure out the probabilities $P(N = k)$ for $k = 0, 1, 2, 3, \dots, n$.

We define the *binomial probability distribution* by

$$b(k, n; p) = P(N = k)$$

Thus, $b(k, n; p)$ is the probability of exactly k successes in n repetitions of the same experiment, where p is the probability of a success in a single experiment. The notation $b(k, n; p)$ keeps track of all the values that we need: k (the number of successes), n (the number of experiments), and p (the probability of a success).

The main objective of this section is to find the formula for $b(k, n; p)$.

Example 10.5   First Step in Answering Questions (i)–(iii) from Example 10.4

To calculate the probability that we flip exactly seven tails in ten flips of a coin we need to know the probability of seven successes in ten experiments, given that the probability of a success is 0.5. Thus, we need to figure out $b(7, 10; 0.5)$.

"Eight virus-free months in a year" means 4 months during which the virus is present. So we are looking for the probability of 4 successes in 12 trials, knowing that the probability of success in each trial is 0.2; i.e., we are looking for $b(4, 12; 0.2)$.

Question (iii): the probability that archeologists find no important artifacts in a month (i.e., zero successes in 30 days) is $b(0, 30; 0.015)$. The probability that on 1 day within a month they find an important artifact is $b(1, 30; 0.015)$. Thus, the probability of finding important artifacts more often than 1 day in a month is (complementary event!) $1 - b(0, 30; 0.015) - b(1, 30; 0.015)$.

Now we work on calculating $b(k, n; p)$.

In the case $n = 1$ (one experiment), there are two outcomes, and their probabilities are recorded in Table 10.3.

Table 10.3

k	$P(N = k)$	$b(k, n; p)$
0	$1 - p$	$b(0, 1; p) = 1 - p$
1	p	$b(1, 1; p) = p$

In words: the probability of success is $b(1, 1; p) = p$ and the probability of no-success is $b(0, 1; p) = 1 - p$. This is not new information, we just restated the assumptions.

Assume that the experiment is repeated twice ($n = 2$). There are four outcomes: success in the first experiment and success in the second experiment (we label this outcome 11), success in the first experiment and no-success in the second experiment (10), no-success in the first experiment and success in the second experiment (01), and no-success in the first experiment and no-success in the second experiment (00). Now the probabilities:

$P(11) = P(\text{success in the first experiment and success in the second experiment})$

$\qquad = P(\text{success in the first experiment})P(\text{success in the second experiment})$

$\qquad = pp = p^2$

Recall that the outcomes of the two experiments are assumed to be independent. That is why we calculated the probability $P(11)$ as the product of the probabilities. We continue in the same way:

$P(10) = P(1 \text{ in the first experiment and 0 in the second experiment})$

$\qquad = P(1 \text{ in the first experiment})P(0 \text{ in the second experiment})$

$\qquad = p(1 - p)$

and

$P(01) = P(0 \text{ in the first experiment and 1 in the second experiment})$

$\qquad = P(0 \text{ in the first experiment})P(1 \text{ in the second experiment})$

$\qquad = (1 - p)p = p(1 - p)$

Finally,

$P(00) = P(0 \text{ in the first experiment and 0 in the second experiment})$

$\qquad = P(0 \text{ in the first experiment})P(0 \text{ in the second experiment})$

$\qquad = (1 - p)(1 - p) = (1 - p)^2$

Recall that N counts the number of successes. Thus, $P(N = 2) = P(11) = p^2$ and $P(N = 0) = P(00) = (1 - p)^2$. Since the outcomes 10 and 01 are mutually exclusive,

$P(N = 1) = P(10 \text{ or } 01) = P(10) + P(01) = p(1 - p) + p(1 - p) = 2p(1 - p)$

Table 10.4 summarizes the probability distribution when $n = 2$.

Table 10.4

k	$P(N = k)$	$b(k, n; p)$
0	$(1 - p)^2$	$b(0, 2; p) = (1 - p)^2$
1	$2p(1 - p)$	$b(1, 2; p) = 2p(1 - p)$
2	p^2	$b(2, 2; p) = p^2$

Next, we repeat the experiment three times ($n = 3$). There are eight possible outcomes $\{111, 011, 101, 110, 001, 010, 100, 000\}$. Using independence, we calculate the probabilities as the products of probabilities:

$$P(111) = P(1 \text{ in the first experiment and } 1 \text{ in the second experiment}$$
$$\text{and } 1 \text{ in the third experiment})$$
$$= P(1 \text{ in the first experiment})P(1 \text{ in the second experiment})$$
$$P(1 \text{ in the third experiment})$$
$$= ppp = p^3$$

To save space, we omit the words since the meaning is clear:

$$P(011) = P(0)P(1)P(1) = (1-p)pp = p^2(1-p)$$

Likewise,

$$P(101) = P(110) = p^2(1-p)$$
$$P(001) = P(010) = P(001) = p(1-p)^2$$

and

$$P(000) = (1-p)^3$$

Therefore, $P(N = 3) = P(111) = p^3$ and $P(N = 0) = P(000) = (1-p)^3$. To compute the remaing two probabilities, we use the mutual exclusivity of events:

$$P(N = 2) = P(011 \text{ or } 101 \text{ or } 110) = P(011) + P(101) + P(110) = 3p^2(1-p)$$

and

$$P(N = 1) = P(001 \text{ or } 010 \text{ or } 100) = P(001) + P(010) + P(100) = 3p(1-p)^2$$

The probability distribution is given in Table 10.5.

Table 10.5

k	$P(N = k)$	$b(k, n; p)$
0	$(1-p)^3$	$b(0, 3; p) = (1-p)^3$
1	$3p(1-p)^2$	$b(1, 3; p) = 3p(1-p)^2$
2	$3p^2(1-p)$	$b(2, 3; p) = 3p^2(1-p)$
3	p^3	$b(3, 3; p) = p^3$

Example 10.6   **Probability of Finding Artifacts; Example 10.3**

The probability that archeologists find an important artifact on any given day is 0.015. Find and interpret $b(1, 2; 0.015)$, $b(0, 3; 0.015)$, and $b(2, 3; 0.015)$.

▶ It is given that the probability of success is $p = 0.015$. The number $b(1, 2; 0.015)$ is the probability that in 1 of 2 days the archeologists find an important artifact. Using Table 10.4 we calculate

$$b(1, 2; 0.015) = 2(0.015)(1 - 0.015) = 0.02955$$

So, the probability is a bit less than 3%.

The number (calculated from Table 10.5)

$$b(0, 3; 0.015) = (1 - 0.015)^3 \approx 0.95567$$

is the probability of zero successes in three repetitions of the experiment. It represents the probability that out of 3 days, no important objects are found on either day. The number

$$b(2, 3; 0.015) = 3(0.015)^2(1 - 0.015) \approx 0.00066$$

is the probability of two successes in three trials, given that the probability of success is 0.015. Thus, the chance that the archeologists find significant objects on exactly 2 days out of 3 days is 0.00066.

Example 10.7   Histogram for a Coin-Tossing Experiment

Consider tossing a fair coin three times in a row, and let the random variable N count the number of tails; so success is flipping tails, and the probability of success is 0.5.

Using Table 10.5, we calculate the probability distribution of N:

$$P(N = 0) = b(0, 3; 0.5) = (1 - 0.5)^3 = 0.125$$
$$P(N = 1) = b(1, 3; 0.5) = 3(0.5)(1 - 0.5)^2 = 0.375$$
$$P(N = 2) = b(2, 3; 0.5) = 3(0.5)^2(1 - 0.5) = 0.375$$
$$P(N = 3) = b(3, 3; 0.5) = (0.5)^3 = 0.125$$

For instance, the probability of flipping exactly two tails in three trials is $P(N = 2) = 0.375$. The probability of flipping no tails in three trials is $P(N = 0) = 0.125$. The histogram for N is drawn in Figure 10.1.

FIGURE 10.1

The histogram for Example 10.7

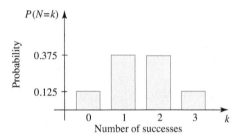

Now we calculate the probability distribution in general, for n repetitions of a Bernoulli experiment. The analysis that we conducted for small values of n suggests that we have to do two things: calculate the probability of each possible outcome, and combine the outcomes that have the same number of successes. For instance, in the case $n = 3$ there are three possible outcomes when $N = 2$ (two successes): 110, 101, and 011. The probability of each occurring is $p^2(1 - p)$, and so $P(N = 2) = 3p^2(1 - p)$.

An outcome that results in exactly k successes (with probability of success p) in n experiments must have $n - k$ no-successes (with probability of no-success $1 - p$). The probability of that particular outcome occurring is

$$\text{(probability of success)}^k \text{(probability of no-success)}^{n-k} = p^k(1 - p)^{n-k}$$

Therefore, the probability of k successes in n experiments is

$b(k, n; p) = $ (number of ways of obtaining k successes in n experiments)$\cdot$

(probability of *one* outcome with k successes in n experiments)

$$= C(n, k) \cdot p^k(1 - p)^{n-k} \tag{10.1}$$

Once we find $C(n, k)$, we are done.

We already know some values: from Table 10.3, $C(1, 0) = 1$, and $C(1, 1) = 1$ (this time, we are looking for the *coefficients* in $b(k, n; p)$). From Table 10.4, $C(2, 0) = 1$, $C(2, 1) = 2$, and $C(2, 2) = 1$. From Table 10.5, $C(3, 0) = 1$, $C(3, 1) = 3$, $C(3, 2) = 3$, and $C(3, 3) = 1$. Look familiar?

If we arrange these values in a triangle, we get

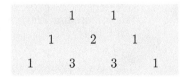

Pascal's triangle! So the numbers $C(n, k)$ seem to be the same as the coefficients in the expansion of $(a + b)^2$, $(a + b)^3$, etc. (these numbers are called *binomial coefficients*).

If this were really so, continuing to build Pascal's triangle

we would obtain $C(4, 0) = 1, C(4, 1) = 4, C(4, 2) = 6, C(4, 3) = 4$, and $C(4, 4) = 1$. Let's check.

$C(4, 0)$ is the number of outcomes with no success in four experiments. There is only one such outcome, 0000; thus, $C(4, 0) = 1$. To find $C(4, 1)$ we need to count the number of ways of obtaining one success in four experiments. Listing all possibilities—1000, 0100, 0010, 0001—we see that $C(4, 1) = 4$. To calculate $C(4, 2)$ we list all outcomes that contain two successes: 1100, 1010, 1001, 0110, 0101, 0011. Thus, $C(4, 2) = 6$, and so on.

But how can we calculate $C(4, 2)$, and $C(n, k)$ in general, without listing all possibilities? To answer this question, we need to learn how to *count* things.

Counting

We discuss several relevant situations that involve counting possibilities.

Example 10.8   Counting: Dividing a Sample into Groups

For a study on the growth of ears in dogs, we organize a sample of dogs into groups. First, we separate female dogs from male dogs. Then, within each group, we divide the dogs according to their body size into small, medium, and large, and then further into short-eared and long-eared. How many groups of dogs are we going to have?

▶ A tree diagram helps; see Figure 10.2.

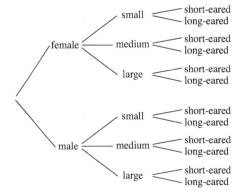

FIGURE 10.2

Conuting: tree diagram

There will be 2 (male or female) times 3 (small or medium or large) times 2 (short-eared or long-eared) = 12 groups. █

This is an example of the *multiplication principle*. In general, if task 1 involves selecting one of n_1 options, task 2 involves selecting one of n_2 options, ..., task k involves selecting one of n_k options, then the multiplication principle says that the total number of outcomes is the product $n_1 n_2 \cdots n_k$.

A coffee shop offers three different sizes of coffee (small, medium, large), six different flavours (Colombian, Cuban, Indonesian, Kenyan, French Vanilla, Irish Cream), four options for milk (no milk, 2% milk, 5% milk, cream), and two options for sugar (sugar or no sugar). There is a total of $3 \cdot 6 \cdot 4 \cdot 2 = 144$ different coffees we can order.

Example 10.9   Counting: Arrangements, Order Matters

In how many different ways can five people be seated in a row on a bench?

▶ Think of an empty bench, and the ways in which the five people can be seated on it, one by one. There are five choices for the leftmost position on the bench. The next position to the right can be occupied by any of the remaining four people, and the next position by any of the remaining three people. There are only two people left unseated, so there are two options for the fourth position from the left, and only one for the fifth. By the multiplication principle, there are $5 \cdot 4 \cdot 3 \cdot 2 \cdot 1 = 120$ different seating arrangements for the five people to sit on the bench. ▲

We introduce a piece of notation. For a positive integer n, we define $n!$ (read "n factorial") by

$$n! = 1 \cdot 2 \cdot 3 \cdots (n - 1) \cdot n$$

In words, $n!$ is the product of all integers from 1 to n. Thus,

$$1! = 1$$
$$2! = 1 \cdot 2 = 2$$
$$3! = 1 \cdot 2 \cdot 3 = 6$$
$$4! = 1 \cdot 2 \cdot 3 \cdot 4 = 24$$
$$5! = 1 \cdot 2 \cdot 3 \cdot 4 \cdot 5 = 120$$

and so on. Factorials grow large very quickly: for instance, $10! = 3,628,800$, $20! = 2,432,902,008,176,640,000$, and $100!$ is larger than 10^{157}. There are situations where taking $0! = 1$ makes sense, so we add it to the definition of the factorial.

We now formalize the reasoning from Example 10.9.

Definition 32 Permutation

A *permutation* on a set of n distinct elements is an ordering of those elements. ▲

To order a set we specify which element is the first, which is the second, and so on. In other words, order matters.

Consider the set of six letters {A, B, C, D, E, F}. The orderings ACDBFE, EFACBD, and CABDEF are three permutations. The permutation CADBEF is different from the permutation ACDBEF.

The number of ways in which we can order n distinct objects is $n!$. Thus, there are $n!$ permutations of a given set of n distinct elements.

Next, we consider a situation where the order of the objects is not relevant.

Example 10.10   Counting: Arrangements, Order Does Not Matter

In how many different ways can we select a group of three people from a group of five people, A, B, C, D, and E?

▶ Think of a bench that has space for three people only. Again, we have five choices for the leftmost position, four choices for the position to the right of it, and three for the remaining position. Thus, there are $5 \cdot 4 \cdot 3$ possibilities, *if* we care about the order (in which case we count ABE, AEB, BEA and other permutations of these three people as distinct possibilities).

But this time, we do not care about the order, so we need to divide $5 \cdot 4 \cdot 3$ by the number of times each subset of three people is counted in. There are three people, so each subset is counted in $3! = 6$ times. We conclude that the total number of ways of picking a group of three people from a group of five is

$$\frac{5 \cdot 4 \cdot 3}{3!} = \frac{60}{6} = 10$$

For the record, we list them all here: ABC, ABD, ABE, ACD, ACE, ADE, BCD, BCE, BDE, CDE.

We introduce the notation

$$\binom{5}{3} = \frac{5 \cdot 4 \cdot 3}{3!}$$

The symbol on the left is read "5 choose 3" (as in "given 5 objects choose 3").

In general, the number of ways in which we can pick k objects from a group of n objects (disregarding order) is given by the *binomial coefficient*

$$\binom{n}{k} = \frac{\text{number of ways to select an } ordered \text{ set of } k \text{ objects}}{\text{number of times each set of } k \text{ objects is counted in}}$$

Since we have n choices for the first object, $n - 1$ choices for the second object, $n - 2$ choices for the third object, ..., $n - k + 1$ choices for the kth object, the total number of ways to select an ordered set of k objects is $n(n-1)(n-2) \cdots (n-k+1)$. The number of times each set of k objects is counted in is equal to the number of permutations of k elements, which is $k!$. Therefore,

$$\binom{n}{k} = \frac{n(n-1)(n-2) \cdots (n-k+1)}{k!} \tag{10.2}$$

We are done. Now think of "object" as "success." We can pick k successes (objects) out of n experiments (objects) in

$$C(n, k) = \binom{n}{k}$$

ways.

Going back to where we were before our detour into counting—equation (10.1)—we get that the probability distribution of the binomial variable is

$$b(k, n; p) = \binom{n}{k} p^k (1 - p)^{n-k}$$

In words, the number $b(k, n; p)$ is the probability of k successes in n repetitions of the same Bernoulli experiment, given that the probability of success is p.

Next, we simplify the fraction on the right side in (10.2). Multiply and divide by $1 \cdot 2 \cdot 3 \cdots (n - k)$

$$\binom{n}{k} = \frac{n(n-1)(n-2) \cdots (n-k+1)}{k!}$$

$$= \frac{n(n-1)(n-2) \cdots (n-k+1)}{k!} \cdot \frac{(n-k) \cdots 3 \cdot 2 \cdot 1}{1 \cdot 2 \cdot 3 \cdots (n-k)}$$

$$= \frac{n!}{k!(n-k)!} \tag{10.3}$$

Thus, we expressed the binomial coefficient $\binom{n}{k}$ using factorials only. For instance,

$$\binom{4}{2} = \frac{4!}{2!(4-2)!} = \frac{4!}{2!2!} = \frac{24}{4} = 6$$

We summarize the most important results in this section in the statements of the following two theorems.

Theorem 10 The Number of Successes in Repeated Bernoulli Experiments

The number of ways to choose a group of k objects out of n objects or, equivalently, k successes in n repetitions of a Bernoulli experiment is equal to

$$\binom{n}{k} = \frac{n!}{k!(n-k)!}$$

Recall that we defined $0! = 1$. Thus, using the formula from Theorem 10,

$$\binom{n}{0} = \frac{n!}{0!(n-0)!} = \frac{n!}{1 \cdot n!} = 1$$

As well,

$$\binom{n}{1} = \frac{n!}{1!(n-1)!} = \frac{n!}{(n-1)!} = \frac{1 \cdot 2 \cdot 3 \cdots (n-1)n}{1 \cdot 2 \cdot 3 \cdots (n-1)} = n$$

Theorem 11 Probability Distribution of the Binomial Variable

The probability distribution of the binomial variable N is given by

$$P(N = k) = b(k, n; p) = \binom{n}{k} p^k (1-p)^{n-k}$$

where N counts the number of successes in n independent repetitions of the same Bernoulli experiment and p is the probability of success.

We are ready to answer the questions that we asked at the beginning of this section.

Example 10.11   Answers to Questions from Example 10.4

(i) The probability that we obtain exactly 7 tails in 10 flips of a fair coin is

$$b(7, 10; 0.5) = \binom{10}{7} (0.5)^7 (1 - 0.5)^{10-7}$$

Using (10.3) or Theorem 10 we compute

$$\binom{10}{7} = \frac{10!}{7! \, 3!} = \frac{3{,}628{,}800}{5{,}040 \cdot 6} = 120$$

and so

$$b(7, 10; 0.5) = 120 \, (0.5)^7 (0.5)^3 = 120(0.5)^{10} \approx 0.117$$

(ii) Recall that the probability of a virus occurring in a given month (that is what the success is in this case) is 0.2. The probability that there will be exactly eight virus-free months is the same as the probability that the virus will be present during exactly 4 months:

$$b(4, 12; 0.2) = \binom{12}{4} (0.2)^4 (1 - 0.2)^{12-4} = 495 \, (0.2)^4 (0.8)^8 \approx 0.133$$

since

$$\binom{12}{4} = \frac{12!}{4! \, 8!} = \frac{479{,}001{,}600}{24 \cdot 40{,}320} = 495$$

Alternatively, we use (10.2) and cancel the common factors in the fraction:

$$\binom{12}{4} = \frac{12 \cdot 11 \cdot 10 \cdot 9}{1 \cdot 2 \cdot 3 \cdot 4} = \frac{11 \cdot 5 \cdot 9}{1} = 495$$

At the end of the section we will say a bit more about calculations of binomial coefficients that involve large numbers.

(iii) The number

$$b(0, 30; 0.015) = \binom{30}{0} (0.015)^0 (1 - 0.015)^{30-0} = 0.985^{30} \approx 0.635$$

is the probability of finding no significant artifacts on any day within a month, assuming that a month has 30 days. (In the above calculation we used the fact that $\binom{30}{0} = 1$; see the note that follows Theorem 10.) The number

$$b(1, 30; 0.015) = \binom{30}{1}(0.015)^1(1 - 0.015)^{30-1} = 30 \cdot (0.015)(0.985)^{29} \approx 0.290$$

is the probability of finding a significant object on exactly 1 day within a month. Therefore, the probability of finding important artifacts more often than 1 day in a month is $1 - b(0, 30; 0.015) - b(1, 30; 0.015) \approx 1 - 0.635 - 0.290 = 0.075$. ▲

The Mean and Variance of the Binomial Distribution

The probability distribution of a Bernoulli random variable B is $P(B = 0) = 1 - p$ and $P(B = 1) = p$. Thus,

$$E(B) = 0 \cdot (1 - p) + 1 \cdot p = p$$

and

$$\begin{aligned}\text{var}(B) &= (0 - p)^2 \cdot (1 - p) + (1 - p)^2 \cdot p \\ &= p^2(1 - p) + (1 - p)^2 p \\ &= p(1 - p)\left[p + (1 - p)\right] \\ &= p(1 - p)\end{aligned}$$

Alternatively, we use the formula $\text{var}(B) = E(B^2) - (E(B))^2$. Note that $P(B^2 = 0) = 1 - p$ and $P(B^2 = 1) = p$ and thus

$$\text{var}(B) = [0 \cdot (1 - p) + 1 \cdot p] - (p)^2 = p - p^2 = p(1 - p)$$

Let

$$N = B_1 + B_2 + \cdots + B_n$$

where B_i, $i = 1, 2, \ldots, n$, are independent Bernoulli random variables (and so N is the binomial distribution). Using Theorem 7 in Section 7 for the mean and Theorem 9 in Section 9 for the variance, we get

$$E(N) = E(B_1) + E(B_2) + \cdots + E(B_n) = np \tag{10.4}$$

and

$$\text{var}(N) = \text{var}(B_1) + \text{var}(B_2) + \cdots + \text{var}(B_n) = np(1 - p) \tag{10.5}$$

Example 10.12 **Mean and Variance in Flipping Coins**

Let N be the number of tails in n tosses of a fair coin. Since $p = 0.5$, the expected value is $E(N) = 0.5n$. Thus, if we toss a coin $n = 100$ times, the expected number of tails is $E(N) = 100(0.5) = 50$.

The histogram of N is drawn in Figure 10.3.

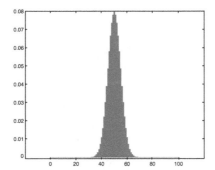

FIGURE 10.3

Histogram of the binomial distribution N

The variance of N is

$$\text{var}(N) = n(0.5)(1 - 0.5) = 0.25n$$

When $n = 100$, then $\text{var}(N) = 0.25(100) = 25$, and the standard deviation is $\sigma = \sqrt{\text{var}(N)} = 5$.

In Section 14 we study bell-shaped distributions in detail. One fact we will discover is that most of a bell-shaped distribution lies within two standard deviations of the mean. The distribution here looks bell-shaped, so we conclude that the number of tails in 100 tosses of a coin will very likely be between $50 - 2 \cdot 5 = 40$ and $50 + 2 \cdot 5 = 60$.

Example 10.13 **Genetics of Green-Eyed Kittens**

Consider two alleles R and G, where R is the dominant allele (red eyes in cats) and G is the recessive allele (green eyes in cats). Assume that there are 30 kittens, all from RG parents, and that the genetic makeup of a kitten is independent of other births. What is the probability that at most three kittens have green eyes?

▶ Define the Bernoulli experiment B by

$$B = \begin{cases} 1 & \text{a kitten has green eyes (success)} \\ 0 & \text{a kitten has red eyes} \end{cases}$$

Of the four genotypes RR, RG, GR, and GG that a kitten can inherit, green eyes results from the trait of GG only. Thus the probability of success is $p = P(B = 1) = 1/4 = 0.25$. The experiment is repeated 30 times, and we declare N to be the number of kittens with green eyes.

The event "at most three kittens have green eyes" is the union of four mutually exclusive events "no kittens have green eyes," "exactly one kitten has green eyes," "exactly two kittens have green eyes," and "exactly three kittens have green eyes." Thus

$$P(N \leq 3) = P(N = 0) + P(N = 1) + P(N = 2) + P(N = 3)$$
$$= b(0, 30; 0.25) + b(1, 30; 0.25) + b(2, 30; 0.25) + b(3, 30; 0.25)$$
$$= \binom{30}{0}(0.25)^0(0.75)^{30} + \binom{30}{1}(0.25)^1(0.75)^{29} + \binom{30}{2}(0.25)^2(0.75)^{28}$$
$$+ \binom{30}{3}(0.25)^3(0.75)^{27}$$
$$\approx 0.000179 + 0.001786 + 0.008631 + 0.026853 = 0.037449$$

i.e., about 3.75%.

Note that zero, one, two, or three kittens are at the lower end of the distribution. For instance, the probability that exactly seven kittens have green eyes is

$$P(N = 7) = b(7, 30; 0.25) = \binom{30}{7}(0.25)^7(0.75)^{23} \approx 0.166236$$

i.e., a bit over 16.6%.

In Figure 10.4 we drew the histogram, showing that the distribution is skewed towards the values $N = 7$ and $N = 8$. Not surprising—the expected number of kittens with green eyes is

$$E(N) = np = 30(0.25) = 7.5$$

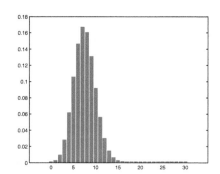

FIGURE 10.4

Histogram of the number of kittens with green eyes

Example 10.14   Occurrence of a Genetic Disorder

Assume that the probability that a child inherits a genetic disorder (such as certain forms of lactose intolerance) from her/his parents is 0.035. A couple decides to have four children. What is the probability that

(i) No children will inherit the disorder.

(ii) At least one child will inherit the disorder.

▶ Define the Bernoulli experiment B by

$$B = \begin{cases} 1 & \text{child inherits the disorder (success)} \\ 0 & \text{child does not inherit the disorder} \end{cases}$$

It is given that $p = P(\text{child inherits the disorder}) = 0.035$. Denote by N the number of children who will inherit the disorder. We assume that the four births are independent events.

(i) The probability is given by (no successes in four trials)

$$P(N = 0) = b(0, 4; 0.035) = \binom{4}{0}(0.035)^0(0.965)^4 = 0.965^4 \approx 0.8672$$

or about 86.7%.

(ii) One way to answer this question is to calculate $P(N = 1) = P(\text{exactly one child will inherit the disorder})$, $P(N = 2) = P(\text{exactly two children will inherit the disorder})$, $P(N = 3) = P(\text{exactly three children will inherit the disorder})$, and $P(N = 4) = P(\text{all four children will inherit the disorder})$ and add up the four probabilities (since the four events are mutually exclusive).

There is an easier way: the complementary event of "at least one child will inherit the disorder" is "no children will inherit the disorder," the probability of which we calculated in (i). Thus,

$$P(\text{at least one child will inherit the disorder})$$
$$= 1 - P(\text{no children will inherit the disorder})$$
$$= 1 - 0.8672 = 0.1328.$$

Example 10.15   Expectation and Variance for the Virus Occurrence from Example 10.2

The occurrence of the virus within a given month is a Bernoulli experiment that can be described by

$$V = \begin{cases} 1 & \text{virus is present (success)} \\ 0 & \text{virus is absent} \end{cases}$$

where $P(V = 1) = 0.2$ and $P(V = 0) = 0.8$. Thus, $E(V) = p = 0.2$ and $\text{var}(V) = p(1 - p) = (0.2)(0.8) = 0.16$. Let the random variable M count the number of months in a 10-year period during which the virus is present in a population. Its histogram is shown in Figure 10.5.

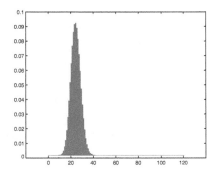

FIGURE 10.5

Histogram of the occurrence
of the virus

The fact that the expected value is $E(M) = np = (120)(0.2) = 24$ explains why the histogram is skewed toward lower values. In words, it is expected that, in a 10-year period, there are 24 months during which the virus is present. The variance of M is

$$\text{var}(M) = np(1 - p) = 120(0.2)(0.8) = 19.2$$

and the standard deviation is $\sigma = \sqrt{19.2} \approx 4.38$.

Appendix: Calculating Factorials and Binomial Coefficients

Calculating binomial coefficients and the probabilities associated with the binomial distribution can be quite challenging.

Back to Example 10.15: in order to figure out the probability that in 10 years there will be exactly 36 months during which the virus is present, we need to find the numeric value of the expression

$$b(36, 120; 0.2) = \binom{120}{36} (0.2)^{36}(0.8)^{84} \tag{10.6}$$

We have two major problems: one is to calculate the binomial coefficient

$$\binom{120}{36} = \frac{120!}{36! \, 84!}$$

since the numbers involved are huge: $120!$ is of the order of magnitude of 10^{198}, $36! \approx 3.720 \cdot 10^{41}$ and $84! \approx 3.314 \cdot 10^{126}$. The other problem is multiplication of very large numbers by very small numbers. The two terms coming from the probabilities are $0.2^{36} \approx 6.872 \cdot 10^{-26}$ and $0.8^{84} \approx 7.237 \cdot 10^{-9}$. Mathematical software can deal with these types of calculations. Using Maple, we find that

$$\binom{120}{36} = 5,425,936,737,585,192,491,355,436,069,690 \approx 5.426 \cdot 10^{30}$$

and

$$b(36, 120; 0.2) = \binom{120}{36} (0.2)^{36}(0.8)^{84} \approx 0.002698$$

There is a way to approximate (10.6) using *Stirling's approximaton* for factorials:

$$n! \approx \sqrt{2\pi n} \left(\frac{n}{e}\right)^n \tag{10.7}$$

In Table 10.6 we show several values of n and (avoiding listing the actual and approximated values of $n!$) the value of the fraction

$$\text{comparison} = \frac{\text{true value of } n!}{\text{Stirling's approximation of } n!}$$

as a measure of how close the approximation is.

NEL

Table 10.6

n	Comparison (true/Stirling's approximation)
20	1.004175
50	1.001668
100	1.000834
200	1.000417

Using Stirling's approximation, we write

$$\binom{120}{36} = \frac{120!}{36!\,84!} \approx \frac{\sqrt{240\pi}\,(120/e)^{120}}{\sqrt{72\pi}\,(36/e)^{36}\,\sqrt{168\pi}\,(84/e)^{84}}$$

It's messy, but it can be done on a hand calculator. As well, we can use logarithms (see Exercise 37).

Summary   A **Bernoulli random variable** is a discrete random variable that takes on two values; one, called the success, has probability p. The number of successes in n independent repetitions of a Bernoulli experiment is given by the **binomial distribution**. In particular, $b(k, n; p)$ gives the probability of exactly k successes in n repetitions of the same Bernoulli experiment. Finding a probability distribution of a binomial random variable involves counting arguments. The mean of the binomial distribution $b(k, n; p)$ is np, and the variance is $np(1-p)$. A number of applications that we explored in this section (such as distribution of alleles, occurrence of a genetic disorder, and dynamics of a virus occurrence in a population) were modelled by the binomial random variable.

10   Exercises

1. There is no time to interview all 20 job applicants (10 women and 10 men), so it has been decided that only 8 will be interviewed. One person at a time is selected randomly and called for the interview. Let X = "number of males interviewed." Is X a binomial variable? Why or why not?

2. A shipment of 2,000 containers has arrived at the port of Vancouver. As part of the customs inspection, a container is selected at random and checked for contraband (say, illegal drugs). Then, of the remaining 1,999 containers, another container is selected at random and checked for illegal drugs. This routine is repeated 25 times. Assume that all containers have the same probability of carrying illegal drugs, and let X = "number of containers that contain illegal drugs." Is X a binomial variable? Why or why not?

3. A northern goshawk preys on fish 40% of the time and on small mammals 60% of the time. A group of ten goshawks (assumed to be acting independently) hunt within the same region. Let X = "number of small mammals captured." Is X a binomial variable? Why or why not?

4. You should have discovered that X in Exercises 1 and 2 is not a binomial variable. Suggest a modification to the routines that will make X a binomial variable. Do these new routines make sense in reality?

5. Suppose that the probability of a success is 0.3. Make a complete list of ways in which we can obtain exactly two successes in four trials. Based on your list, find the probability of obtaining exactly two successes in four independent trials. Compare with the formula for the binomial distribution.

6. Suppose that the probability of a success is 0.3. Make a complete list of ways in which we can obtain exactly two successes in five trials. Based on your list, find the probability of obtaining exactly two successes in five independent trials. Compare with the formula for the binomial distribution.

▽ 7–14 ▪ Find each quantity and explain what it represents.

7. $\binom{12}{3}$

8. $\binom{10}{4}$

9. $C(8,0)$

10. $C(13,1)$

11. $b(1,4;0.6)$

12. $b(3,4;0.6)$

◣ 13. $b(1,7;0.2)$

14. $b(1,7;0.8)$

▽ 15–18 ▪ Express each quantity in the form of a binomial coefficient and find its value.

15. The number of ways we can obtain three tails in eight tosses of a coin.

16. The number of ways we can obtain six heads in seven tosses of a coin.

17. The number of ways of selecting a team of 4 students from a group of 20 students.

◣ 18. The number of ways of selecting a team of 11 players from a group of 14 available players.

▽ 19–22 ▪ A random variable X is distributed binomially with probability of success $p = 0.6$. Using the notation $b(k,n;p)$ for the probabilities associated with the binomial distribution, say what you would need to calculate to answer each question.

19. Find the probability of at least three successes in five trials.

20. Find the probability of at most three successes in eight trials.

21. Find the probability that there are no more than 9 and no less than 5 successes in 25 trials.

◣ 22. Find the probability that there are more than 2 successes in 50 trials.

23. Explain why (10.3) implies that $\binom{n}{k} = \binom{n}{n-k}$. Use this fact to calculate $\binom{22}{20}$.

24. Using the formula in Exercise 23, calculate $\binom{50}{47}$.

▽ 25–28 ▪ The random variable X is binomially distributed with parameters n (number of trials) and p (probability of success). In each case:

(a) Find the probability distribution of X.

(b) Draw a histogram of X.

(c) Find the mean and the variance of X using the probabilities you found in (a).

(d) Find the mean and the variance using formulas (10.4) and (10.5) and compare with your answers to (c).

25. $n = 2$, $p = 0.4$

26. $n = 3$, $p = 0.4$

◣ 27. $n = 4$, $p = 0.4$

28. $n = 5$, $p = 0.5$

29. A box of chocolates contains 20 chocolates. The probability that any one chocolate has a hazelnut is 0.03.

(a) What is the expected number of chocolates with a hazelnut per box?

(b) What is the probability that there are no chocolates with a hazelnut in one box of chocolates?

(c) You buy 15 boxes of chocolates. What is the expected number of boxes that contain no chocolates with a hazelnut?

30. About 10% of Canadians have latent TB (tuberculosis) infection (i.e., they have been infected, are not infectious, but can develop tuberculosis at some point in their lives). [Source: Canadian Institute for Health Information, Canadian Lung Association, Health Canada Statistics Canada, *Respiratory Disease in Canada,* September 2001. Available at www.phac-aspc.gc.ca/publicat/rdc-mrc01/.]

    (a) What is the expected number of people with latent TB infection in Winnipeg, Manitoba (population 633 thousand)? Define the binomial variable involved, say what constitutes a success, and state the probability of success.

    (b) What is the probability that in a randomly chosen sample of ten people in Winnipeg nobody has a latent TB infection?

31. It has been determined that 15% of all tomato plants in a greenhouse have been infested with hornworms. You pick ten plants at random. What is the probability that none of them have been infested with hornworms?

32. Suppose that dogs with genotype SS and SC have straight hair, and those with genotype CC have curly hair (so S is the dominant allele). Two SC parents have eight puppies. What is the probability that exactly four puppies have curly hair?

33. Suppose that dogs with genotype SS and SC have straight hair, and those with genotype CC have curly hair (so S is the dominant allele). Two SC parents have eight puppies. What is the expected number of puppies with curly hair? Call that number n_c. What is the probability that exactly n_c puppies have curly hair?

34. Suppose that the alleles responsible for the growth of a trout are additive in the sense that trout with genotype LL are long, those with genotype LS are of medium length, and those with genotype SS are short. A crossing of two LS trout produced 12 offspring.

    (a) What is the expected value of the number of long offspring?

    (b) Find the probability that there are at least two long offspring.

35. Suppose that the alleles responsible for the growth of a trout are additive in the sense that trout with genotype LL are long, those with genotype LS are of medium length, and those with genotype SS are short. A crossing of two LS trout produced 12 offspring.

    (a) What is the expected value of the number of medium-sized offspring?

    (b) Find the probability that there are at most two medium-sized offspring.

36. Various surveys have found that about 95% of claims that certain products are "green" (or "eco-friendly" or "organic") are either misleading or not true at all.

    (a) What is the expected number of truly "green" products out of 1,000 products that are claimed to be "green?"

    (b) You buy five products that are claimed to be "green." What is the probability that none of them are truly "green"? What is the probability that all of them are truly "green"?

37. We investigate the approximation of factorials using Stirling's formula.

    (a) Using Stirling's formula, approximate 50! and compare with the true value.

    (b) Compute the logarithm with base 10 of the formula (10.7) and use it to estimate the number of digits in 120!. (Thus, you will check the claim made in the text that 120! is of the order of magnitude of 10^{198}.)

    (c) Approximate $\log_{10}\binom{120}{36}$ using a calculator. Compare with the true value of 30.73447.

In this section we study two discrete distributions that we find often in applications: the **multinomial** and the **geometric distributions.** We start by generalizing the binomial distribution.

The Multinomial Distribution

The binomial distribution is based on repetitions of an experiment (called the Bernoulli experiment or the Bernoulli trial) that has *two* outcomes.

Now consider an experiment that has three possible outcomes, called A_1, A_2, and A_3, whose probabilities are given in Table 11.1. Repeat the experiment ten times, and assume that the experiments are independent of each other. Define random variables N_1 = "number of times event A_1 occurs," N_2 = "number of times event A_2 occurs," and N_3 = "number of times event A_3 occurs." What is the probability that $N_1 = 3$, $N_2 = 5$, and $N_3 = 2$?

Table 11.1

Outcome	Probability
A_1	0.2
A_2	0.3
A_3	0.5

We argue as in the binomial case: the probability that A_1 occurs exactly three times is 0.2^3; the probability that A_2 occurs exactly five times is 0.3^5; and the probability that A_3 occurs exactly two times is 0.5^2. Thus, the probability that *one* event with $N_1 = 3$, $N_2 = 5$, and $N_3 = 2$ occurs is

$$0.2^3 \cdot 0.3^5 \cdot 0.5^2$$

How many events with $N_1 = 3$, $N_2 = 5$, and $N_3 = 2$ are there? Their number is equal to the number of ways in which we can arrange ten numbers into three groups, the first one containing three numbers, the second five numbers, and the third two numbers.

There is a total of 10! arrangements of ten numbers. Note that each choice of three numbers appears in 3! arrangements; each choice of five numbers appears in 5! arrangements, and each choice of two numbers appears in 2! arrangements. Thus, the total number of arrangements is

$$\frac{10!}{3! \cdot 5! \cdot 2!} \tag{11.1}$$

and the probability is (different arrangements are mutually exclusive, so we add up the probabilities)

$$P(N_1 = 3, N_2 = 5, N_3 = 2) = \frac{10!}{3! \cdot 5! \cdot 2!} \, 0.2^3 \cdot 0.3^5 \cdot 0.5^2$$

See Exercises 1 and 2 for practice on counting.

Example 11.1   **Eating Habits of Wolves in Algonquin Park**

Wolf predation can have a considerable impact on the vulnerability and density of prey in an ecosystem. Studying wolves in Algonquin Provincial Park in Ontario, a group of researchers determined the probabilities for the prey of a single wolf; see Table 11.2. [Source: Voigt, D.R., Kolenosky, G.B., & Pimlott, D.H. (1976).

Changes in summer foods of wolves in Central Ontario. *The Journal of Wildlife Management,* 40 (4), 663-668.]

Table 11.2

Prey	Probability
Deer	0.33
Beaver	0.55
Moose	0.05
Other	0.07

Consider the impact on the prey by a group of 80 adult wolves, assumed to act independently.

(a) What is the probability that the 80 wolves will prey on 25 deer, 50 beavers, and 5 moose (and no animals from the "other" group)?

(b) What is the probability that the 80 wolves will prey on 24 deer, 53 beavers, 1 moose, and 2 animals from the "other" group?

▶ Define the random variables N_1 = "number of deer preyed on," N_2 = "number of beavers preyed on," N_3 = "number of moose preyed on," and N_4 = "number of other animals preyed on."

(a) Arguing in the same way as in the introductory example, we conclude that

$$P(N_1 = 25, N_2 = 50, N_3 = 5, N_4 = 0)$$
$$= \frac{80!}{25! \cdot 50! \cdot 5! \cdot 0!} \, 0.33^{25} \cdot 0.55^{50} \cdot 0.05^5 \cdot 0.07^0$$
$$\approx 0.0000378$$

(b) Likewise,

$$P(N_1 = 24, N_2 = 53, N_3 = 1, N_4 = 2)$$
$$= \frac{80!}{24! \cdot 53! \cdot 1! \cdot 2!} \, 0.33^{24} \cdot 0.55^{53} \cdot 0.05^1 \cdot 0.07^2$$
$$\approx 0.0001595$$

Note that the probabilities are very small. The reason is that there are many ways (over 7 million) in which four non-negative numbers N_1, N_2, N_3, and N_4 can be added up to give 80.

The Geometric Distribution

We go back to the repetitions of Bernoulli trials. Recall that a Bernoulli trial (Bernoulli experiment) has two outcomes: success (probability p) and no-success (probability $1-p$). Assume that the trials are independent, and define the random variable

$$X = \text{"number of trials until the first success"}$$

The range of X is the set $1, 2, 3, 4, \ldots$. Thus, X is a discrete random variable whose range is a countably infinite set. (Recall that we have already met countably infinite discrete random variables in Examples 6.4 and 6.5 in Section 6.) What is the probability mass function of X?

The probability $P(X = 1)$ is the probability of success in the first trial, so $P(X = 1) = p$. By independence,

$$P(X = 2) = P(\text{the first trial is no-success and the second trial is success})$$
$$= P(\text{the first trial is no-success})P(\text{the second trial is success})$$
$$= (1 - p)p$$

Likewise,

$$\begin{aligned}
P(X = 3) &= P(\text{no-success then no-success then success}) \\
&= P(\text{no-success})P(\text{no-success})P(\text{success}) \\
&= (1-p)(1-p)p \\
&= (1-p)^2 p
\end{aligned}$$

In general,

$$\begin{aligned}
P(X = k) &= P(\text{no-success in the first } k-1 \text{ trials and success in } k\text{th trial}) \\
&= P(\text{no-success in the first } k-1 \text{ trials})P(\text{success in } k\text{th trial}) \\
&= (1-p)^{k-1} p \qquad\qquad (11.2)
\end{aligned}$$

The formula (11.2) holds for $k = 1, 2, 3, \ldots$. In order to show that it represents a probability mass function, we need to check that

(1) $P(X = k) \geq 0$ for all k.

(2) The sum of all probabilities in (11.2) is 1.

Since $0 \leq p \leq 1$, it follows that $1 - p \geq 0$ and thus $(1-p)^{k-1} \geq 0$ for all k; so (1) is true.

Note that if $p = 0$ (the probability of success is zero), then the random variable X makes no sense (as it describes something that cannot happen). If $p = 1$ then $P(X = 1) = 1$ is the complete probability mass function of X. To avoid these two trivial cases, in what follows we assume that $0 < p < 1$.

For (2), we need to show that the *infinite* sum

$$\sum_{k=1}^{\infty} P(X = k) = 1$$

By definition, the infinite sum (also known as the *series*) is calculated as the limit of finite sums:

$$\sum_{k=1}^{\infty} P(X = k) = \lim_{n \to \infty} \sum_{k=1}^{n} P(X = k)$$

In our case, we need to calculate

$$\begin{aligned}
\sum_{k=1}^{\infty} P(X = k) &= \sum_{k=1}^{\infty} (1-p)^{k-1} p \\
&= p + (1-p)p + (1-p)^2 p + (1-p)^3 p + \cdots \qquad (11.3)
\end{aligned}$$

Sums are usually difficult to calculate explicitly (i.e., using formulas). But we are lucky! We recognize the sum in (11.3) as a *geometric series,* for which we do have a formula.

Recall that if $|q| < 1$, then

$$1 + q + q^2 + q^3 + \cdots = \frac{1}{1-q} \qquad\qquad (11.4)$$

If $|q| \geq 1$, then the sum $1 + q + q^2 + q^3 + \cdots$ is not a real number (we will not worry about it since we will not meet this case). See Exercise 27 for the derivation of the sum of the geometric series formula (11.4).

Using (11.4) with $q = 1 - p < 1$, the sum in (11.3) is calculated to be

$$\begin{aligned}
p + (1-p)p &+ (1-p)^2 p + (1-p)^3 p + \cdots \\
&= p \left[1 + (1-p) + (1-p)^2 + (1-p)^3 + \cdots \right] \\
&= p \, \frac{1}{1 - (1-p)} \\
&= p \, \frac{1}{p} = 1
\end{aligned}$$

Therefore,

$$\sum_{k=1}^{\infty} P(X = k) = \sum_{k=1}^{\infty} (1-p)^{k-1} p = 1$$

and we are done.

Definition 33   The Geometric Distribution

A discrete random variable X is said to be *geometrically distributed with parameter* p if its probability mass function is given by

$$P(X = k) = (1-p)^{k-1} p \qquad (11.5)$$

for $k = 1, 2, 3, \ldots$.

The parameter p is the probability of a success in a single trial. The histograms in Figure 11.1 represent geometric distributions with $p = 0.5$ and $p = 0.15$. The fact that the probabilities decrease means that the largest probability is that of a success happening right away. As we keep repeating the trials, waiting for a success to occur, the probabilities get smaller and smaller.

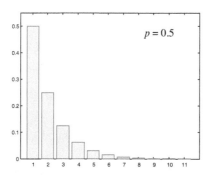
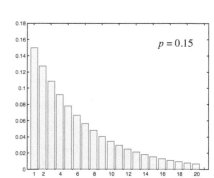

FIGURE 11.1

Geometric distributions

Example 11.2   Molecule Diffusing Out of a Region

The probability that a molecule leaves a fixed region (for instance, a cell) during any given time interval is 0.15; once it leaves, the molecule does not come back. What is the probability that the molecule leaves during the sixth time interval?

▶ Define the random variable $X = $ "time interval during which the molecule leaves the region," and declare the event "molecule leaves the region" to be the success. It is given that the probability of success is $p = 0.15$.

Since X counts the number of trials until the first success, it is geometrically distributed with parameter $p = 0.15$. Thus, the probability that the molecule leaves during the sixth time interval is

$$P(X = 6) = (1-p)^5 \, p = (1 - 0.15)^5 \, 0.15 \approx 0.06656$$

i.e., about 6.7%.

Example 11.3   Molecule Diffusing Out of a Region, II

Continuing with the previous example, what is the probability that the molecule does not leave the region during the first ten time intervals?

▶ The molecule can leave during the 11th, 12th, 13th, ... time intervals. In terms of the random variable X, to answer the question we need to calculate the sum

$$P(X = 11) + P(X = 12) + P(X = 13) + \cdots$$

(which is an infinite sum). Using (11.5), we find
$$P(X = 11) = (1 - p)^{10}\, p = 0.85^{10}\, 0.15$$
$$P(X = 12) = (1 - p)^{11}\, p = 0.85^{11}\, 0.15$$
$$P(X = 13) = (1 - p)^{12}\, p = 0.85^{12}\, 0.15$$

and so on. Thus,
$$P(X = 11) + P(X = 12) + P(X = 13) + \cdots$$
$$= 0.85^{10}\, 0.15 + 0.85^{11}\, 0.15 + 0.85^{12}\, 0.15 + \cdots$$
$$= 0.85^{10}\, 0.15 \left[1 + 0.85 + 0.85^2 + \cdots\right]$$
$$= 0.85^{10}\, 0.15 \cdot \frac{1}{1 - 0.85}$$
$$= 0.85^{10}\, 0.15 \cdot \frac{1}{0.15}$$
$$= 0.85^{10} \approx 0.196874$$

We used the formula for the sum of a geometric series (11.4) with $q = 0.85 < 1$.

This was good practice, but there is a quicker way to do this. The event "molecule does not leave during the first ten time intervals" is the intersection of the events "molecule does not leave during the first time interval," "molecule does not leave during the second time interval," "molecule does not leave during the third time interval," ..., "molecule does not leave during the tenth time interval." Because these ten events are independent, the probability of their intersection is the product of the probabilities. Thus,

$$P(\text{"molecule does not leave during the first ten time intervals"}) = 0.85^{10}$$

There is a way to do it even more quickly: the desired probability is equal to the probability of zero successes in ten trials of a Bernoulli experiment with probability of success equal to $p = 0.15$:

$$b(0, 10; 0.15) = \binom{10}{0} 0.15^0 \, (1 - 0.15)^{10} = 0.85^{10}$$

Let X be a geometric distribution with parameter p, i.e.,
$$P(X = k) = (1 - p)^{k-1} p \tag{11.6}$$

for $k = 1, 2, 3, \ldots$. Arguing as in Example 11.3, we conclude that the probability of no success in the first k trials (i.e., the probability that a success occurs after the kth attempt) is

$$P(X > k) = (1 - p)^k$$

The cumulative distribution function of X, given by $P(X \le k)$, is the probability of success before trial k, or on trial k. It is given by

$$P(X \le k) = 1 - P(X > k) = 1 - (1 - p)^k \tag{11.7}$$

Clearly, $(1 - p)^k \to 0$ as $k \to \infty$ (since $1 - p < 1$). Thus,
$$P(X \le k) \to 1$$

as $k \to \infty$.

In Figure 11.2 we show the cumulative distribution function for the molecule in Examples 11.2 and 11.3 ($p = 0.15$).

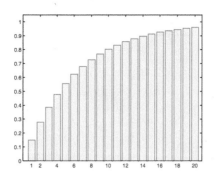

FIGURE 11.2

Cumulative distribution for
the diffusing molecule

Note that

$$P(X = k) = (1 - p)^{k-1} p$$

is a decreasing function of k (see Figure 11.1 and Exercise 28). Thus, the mode of X (that is, the mode of any geometric distribution) is $X = 1$, no matter what the probability of success. The median of X is the integer that is the closest to the value K where the cumulative distribution function is equal to $1/2$, i.e.,

$$1 - (1 - p)^K = 0.5$$
$$(1 - p)^K = 0.5$$
$$K = \frac{\ln 0.5}{\ln(1 - p)}$$

Now we calculate the mean of X. By definition,

$$E(X) = \sum_{k=1}^{\infty} k P(X = k) = \sum_{k=1}^{\infty} k (1 - p)^{k-1} p = p \sum_{k=1}^{\infty} k (1 - p)^{k-1}$$

In Exercise 29(a) we show that

$$\sum_{k=1}^{\infty} k (1 - p)^{k-1} = \frac{1}{p^2}$$

and therefore

$$E(X) = p \frac{1}{p^2} = \frac{1}{p}$$

For the molecule in Example 11.2,

$$E(X) = \frac{1}{0.15} \approx 6.667$$

Thus, on average, the molecule will leave the region during the seventh time interval.

Suppose we keep rolling a die until a 1 appears for the first time. Since the probability of success is $p = 1/6$, the expected value is $E = 1/p = 1/(1/6) = 6$. Thus, on average, we have to roll a die six times until we roll a 1.

To calculate the variance of X, we use $\text{var}(X) = E(X^2) - (E(X))^2$. First we calculate

$$E(X^2) = \sum_{k=1}^{\infty} k^2 P(X = k) = \sum_{k=1}^{\infty} k^2 (1 - p)^{k-1} p = \frac{2 - p}{p^2}$$

(see Exercise 29(b)). Thus,

$$\text{var}(X) = \frac{2 - p}{p^2} - \frac{1}{p^2} = \frac{1 - p}{p^2}$$

When $p = 0.5$,

$$\text{var}(X) = \frac{1 - 0.5}{0.5^2} = 2$$

and the standard deviation is $\sqrt{2} \approx 1.414$. When $p = 0.15$,

$$\text{var}(X) = \frac{1 - 0.15}{0.15^2} \approx 37.778$$

and the standard deviation is 6.146. These calculations agree with the histograms in Figure 11.1: the distribution with $p = 0.15$ is more spread out than the distribution with $p = 0.5$.

Summary   The binomial distribution is based on repetitions of an experiment that has two outcomes. Independent repetitions of an experiment that has more than two outcomes generate the **multinomial distribution.** The probability of the first success in a sequence of repetitions of a Bernoulli experiment is given by the **geometric distribution.** We say that the geometric distribution represents the **waiting time** until the first success. If the probability of success is small, the geometric distribution has a large variance.

11	Exercises

1. Four balls are numbered 1, 2, 3, and 4. Write the answer to each question using factorials, as in (11.1).

   (a) In how many ways can we place the four balls into two boxes so that one box contains one ball and the other box three balls? Write down all possible combinations, to check your answer.

   (b) In how many ways can we place the four balls into two boxes so that each box contains two balls? List all possible combinations.

   (c) In how many ways can we place the four balls into three boxes so that two boxes contain one ball each and the third one contains two balls? List all possible combinations.

2. Five balls are numbered 1, 2, 3, 4, and 5. Write the answer to each question using factorials, as in (11.1).

   (a) In how many ways can we place the five balls into two boxes so that one box contains one ball and the other box four balls? Write down all possible combinations.

   (b) In how many ways can we place the five balls into two boxes so that one box contains two balls and the other three balls? List all possible combinations.

   (c) In how many ways can we place the five balls into three boxes so that two boxes contain one ball each and the third contains three balls? List all possible combinations.

   (d) In how many ways can we place the five balls into three boxes so that two boxes contain two balls each and the third contains one ball? List all possible combinations.

3. Consider the wolf predation described in Example 11.1.

   (a) Write an expression for the probability that the 80 wolves will prey on 10 deer, 70 beavers, no moose, and no animals from the "other" group.

   (b) Write an expression for the probability that the 80 wolves will prey on 60 beavers, 16 animals from the "other" group, and any combination of a total of four deer or moose.

4. Consider the wolf predation described in Example 11.1.

   (a) Write an expression for the probability that the 80 wolves will prey on 5 deer, 65 beavers, 2 moose, and 8 animals from the "other" group.

   (b) Write an expression for the probability that the 80 wolves will prey on 60 beavers, 15 animals from the "other" group, and (of the remaining five animals) more deer than moose.

5. In a crossing of genotype AB parents, three offspring are homozygous of genotype AA, four are homozygous of genotype BB, and two are heterozygous. What is the probability of this event occurring?

6. In a crossing of genotype AB and AB parents, two offspring are of genotype AA, three are of genotype BB, and three are heterozygous. What is the probability of this event occurring?

7. Suppose that the alleles responsible for the growth of a trout are additive in the sense that trout with genotype LL are long, those with genotype LS are of medium length, and those with genotype SS are short. A crossing of two LS trout produced six offspring.

   (a) What is the probability that two offspring are long, two are of medium length, and two are short?

   (b) What is the probability that two offspring are long, and of the remaining four, there are more short than medium-length offspring?

8. Suppose that the alleles responsible for the growth of a trout are additive in the sense that trout with genotype LL are long, those with genotype LS are of medium length, and those with genotype SS are short. A crossing of two LS trout produced six offspring.

   (a) What is the probability that three offspring are long, one is of medium length, and two are short?

   (b) What is the probability that three offspring are short, and of the remaining three, there is at least one long offspring?

9. Assume that A is a normal allele and B is a mutant allele. A pair of mutant alleles BB causes a certain trait (say, attached earlobes). A person with genotype AB is a carrier of the trait but does not exhibit it. A person with genotype AA neither is a carrier nor exhibits the trait. Assume that AB parents decide to have four children. What is the probability that one child will have attached earlobes, two will be carriers, and one will neither be a carrier nor have attached earlobes?

10. Consider the context in Exercise 9. What is the probability that the couple will have exactly two children with attached earlobes?

▽ 11–17 ▪ In each case:

   (a) Compute the required probability.

   (b) Sketch the histogram and shade the area corresponding to the probability in (a).

11. The probability that the first success occurs on the fourth trial, given that the probability of a success in each trial is 0.15.

12. The probability that the first success occurs on the third trial, given that the probability of a success in each trial is 0.8.

13. The probability that the first success occurs on the third trial, given that the probability of a success in each trial is 0.2.

14. The probability that the first success occurs on or after the third trial, given that the probability of a success in each trial is 0.3.

15. The probability that the first success occurs on or after the fourth trial, given that the probability of a success in each trial is 0.6.

16. The probability that the first success occurs on or before the third trial, given that the probability of a success in each trial is 0.3.

17. The probability that the first success occurs on or before the fourth trial, given that the probability of a success in each trial is 0.6.

18. Consider geometric distributions with probabilities of success equal to $p_1 = 0.3$ and $p_2 = 0.7$. Which one has larger the mean? Which is more spread out?

19. Consider geometric distributions with probabilities of success equal to $p_1 = p$ and $p_2 = p/2$. Which distribution is more spread out?

20. Assume a 1:1 sex ratio. A couple continues to have children until they have a girl. Find each probability.

    (a) The second child is a girl.

    (b) Either the second or the third child is a girl.

    (c) A girl is born after four attempts.

21. The mean of a geometric distribution is 5. What is its standard deviation?

22. The mean of a geometric distribution is m. What is its standard deviation?

23. The variance of a geometric distribution is 2. What is its mean?

24. The variance of a geometric distribution is $v > 0$. What is its mean?

25. A gene has a 0.1% chance of mutating each time a cell divides. What is the probability that a gene will mutate during the 20th cell division? What is the probability that it will mutate before or during the 20th cell division?

26. A molecule leaves a cell during each hour with a 45% chance. Find the probability that the molecule leaves during the tenth hour. Find the probability that the molecule leaves before or during the tenth hour.

27. Recall that an infinite sum is calculated as the limit of finite sums. Take a number q such that $|q| < 1$.

    (a) Let $s_n = 1 + q + q^2 + q^3 + \cdots + q^n$. Multiply this equation by q, and subtract what you obtain from the given equation. Solve for s_n to get $s_n = (1 - q^{n+1})/(1 - q)$.

    (b) Calculate the limit of s_n as $n \to \infty$ to show that $1 + q + q^2 + q^3 + \cdots = 1/(1 - q)$.

28. Assume that $0 < p < 1$.

    (a) Explain why the numbers $p, (1 - p)p, (1 - p)^2 p, (1 - p)^3 p, \ldots$ form a decreasing sequence.

    (b) Use calculus to show that $f(k) = (1 - p)^{k-1} p$ is a decreasing function.

29. Assume that $0 < q < 1$.

    (a) Start with

    $$1 + q + q^2 + q^3 + \cdots = \frac{1}{1 - q}$$

    and differentiate both sides with respect to q. Replace q by $1 - p$ to show that

    $$\sum_{k=1}^{\infty} k(1 - p)^{k-1} = \frac{1}{p^2}$$

    (b) Start with

    $$1 + q + q^2 + q^3 + \cdots = \frac{1}{1 - q}$$

    and differentiate both sides with respect to q. Then multiply by q and differentiate both sides with respect to q once again. Replace q by $1 - p$ to obtain

    $$p \sum_{k=1}^{\infty} k^2 (1 - p)^{k-1} = \frac{2 - p}{p^2}$$

| 12 | The Poisson Distribution |

In this section we study one of the most useful discrete distributions, the **Poisson distribution.**

The Poisson Distribution

The Poisson distribution counts the number of events that occur randomly over time or over space. It is one of the most widely used distributions, as it can model a wide range of phenomena, such as the number of mutations in a stretch of a DNA, the number of bacteria per square centimetre of a kitchen countertop, the number of weekly requests for an emergency ultrasound or CT scan, the number of car accidents on a given stretch of highway, the number of stars in a given volume of the sky, or the distribution of pieces of almond in a square of chocolate.

The Poisson distribution describes the number of occurrences of an event in a given finite interval of time or space, assuming that

(1) the occurrences of the event are mutually independent (i.e., one event occurring does not affect the probability of another event occurring), and

(2) the probability of an occurrence is the same for all intervals of the same size.

The "interval" can be an actual time interval, or a unit of length, area, or volume (as we will witness soon). The events that satisfy (1) and (2) are said to form a *Poisson process.*

Definition 34   **The Poisson Distribution**

A random variable X is said to have a *Poisson distribution with parameter* λ if its probability mass function satisfies

$$P(X = k) = e^{-\lambda} \frac{\lambda^k}{k!} \tag{12.1}$$

for $k = 0, 1, 2, \ldots$.

X is a discrete random variable whose range is countably infinite. To denote that X is Poisson-distributed with parameter λ, we write $X \sim \text{Po}(\lambda)$.

Recall that $0! = 1$, $1! = 1$, and $k! = 1 \cdot 2 \cdot 3 \cdots k$, for $k \geq 1$. Since $\lambda > 0$, we conclude that $P(X = k) \geq 0$ for all non-negative k. To show that (12.1) is a probability mass function, we have to to verify that

$$\sum_{k=0}^{\infty} e^{-\lambda} \frac{\lambda^k}{k!} = 1 \tag{12.2}$$

We factor out the term not involving k:

$$\sum_{k=0}^{\infty} e^{-\lambda} \frac{\lambda^k}{k!} = e^{-\lambda} \sum_{k=0}^{\infty} \frac{\lambda^k}{k!}$$

Recall that the Taylor polynomial of $y = e^x$ of order n is given by

$$T_n(x) = 1 + x + \frac{x^2}{2!} + \frac{x^3}{3!} + \cdots + \frac{x^n}{n!} = \sum_{k=0}^{n} \frac{x^k}{k!}$$

Taking the limit of both sides we obtain

$$e^x = \lim_{n \to \infty} T_n(x) = \lim_{n \to \infty} \sum_{k=0}^{n} \frac{x^k}{k!} = \sum_{k=0}^{\infty} \frac{x^k}{k!} \tag{12.3}$$

The last equal sign is due to the definition of an infinite sum. The reasons why the first equal sign in (12.3) is true are beyond the scope of this book.

Using (12.3) with $x = \lambda$, we get

$$e^{-\lambda} \sum_{k=0}^{\infty} \frac{\lambda^k}{k!} = e^{-\lambda} e^{\lambda} = 1$$

So (12.1) is indeed a probability mass function.

In Figure 12.1 we show histograms of Poisson distributions corresponding to four different values of the parameter λ.

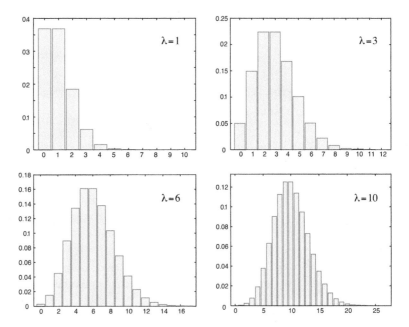

FIGURE 12.1

Poisson distributions

Example 12.1   Bacteria in a Piece of Salami

A piece of salami that was dropped on the floor was found to contain four salmonella bacteria per square centimetre. (As few as ten salmonella bacteria can cause gastroenteritis (stomach flu).) Assuming that the number of salmonella bacteria per square centimetre has a Poisson distribution, find the probability that

(a) There are no salmonella bacteria in a given square centimetre of the piece of salami.

(b) There are at least three bacteria in a given square centimetre of the piece of salami.

▶ Define $X =$ "number of salmonella bacteria in a given square centimetre of the piece of salami." Then $X \sim \text{Po}\,(4)$, i.e., $\lambda = 4$ bacteria per square centimetre.

(a) We are looking for the probability $P(X = 0)$. Substituting $\lambda = 4$ and $k = 0$ into (12.1), we obtain

$$P(X = 0) = e^{-4} \frac{4^0}{0!} = e^{-4} \approx 0.018316$$

Thus, the chance of finding no salmonella bacteria in a randomly chosen square centimetre of the piece of salami is less than 2%.

(b) The probability that there are at least three bacteria in a given square centimetre of the salami is (using complementary events)

$$
\begin{aligned}
P(X \geq 3) &= 1 - P(X < 3) \\
&= 1 - P(X = 0 \text{ or } X = 1 \text{ or } X = 2) \\
&= 1 - (P(X = 0) + P(X = 1) + P(X = 2))
\end{aligned}
$$

The last equal sign is because the three events $X = 0$, $X = 1$, and $X = 2$ are mutually exclusive.

From

$$P(X = 0) + P(X = 1) + P(X = 2) = e^{-4} \frac{4^0}{0!} + e^{-4} \frac{4^1}{1!} + e^{-4} \frac{4^2}{2!}$$

$$= e^{-4} \left(1 + 4 + \frac{4^2}{2} \right)$$

$$= 13e^{-4} \approx 0.238103$$

we obtain

$$P(X \geq 3) = 1 - P(X < 3) \approx 1 - 0.238103 = 0.761897$$

In the previous example, the "interval" was 1 cm^2. In our next example, it is a time interval.

Example 12.2   **The Number of Requests for an Emergency Diagnostic Scan**

The number of emergency requests for an ultrasound or a CT scan during the 9-hour interval from 8 a.m. to 5 p.m. on a weekday in a medical clinic is approximated by the Poisson distribution with parameter $\lambda = 6.56$; units of λ are requests per 9-hour interval. [Source: Vasanawala, S.S., & Desser, T.S. (2005). Accommodation of requests for emergency US and CT: Applications of queueing theory to scheduling of urgent studies. *Radiology*, 235 (1), 244-249.]

What is the probability that, on a given weekday between 8 a.m. and 5 p.m., there will be between 9 and 12 emergency requests for the scans?

▶ Let X denote the number of requests for an emergency scan on a weekday between 8 a.m. and 5 p.m. It is given that $X \sim \text{Po}\,(6.56)$. Thus (using mutual exclusivity of events)

$$P(9 \leq X \leq 12) = P(X = 9 \text{ or } X = 10 \text{ or } X = 11 \text{ or } X = 12)$$

$$= P(X = 9) + P(X = 10) + P(X = 11) + P(X = 12)$$

$$= e^{-6.56} \frac{6.56^9}{9!} + e^{-6.56} \frac{6.56^{10}}{10!} + e^{-6.56} \frac{6.56^{11}}{11!} + e^{-6.56} \frac{6.56^{12}}{12!}$$

$$\approx 0.087781 + 0.057584 + 0.034341 + 0.018773$$

$$= 0.198479$$

Next, we compute the mean and the variance of a Poisson distribution.

The mean of $X \sim \text{Po}\,(\lambda)$ is given by the sum

$$E(X) = \sum_{k=0}^{\infty} kP(X = k) = \sum_{k=0}^{\infty} ke^{-\lambda} \frac{\lambda^k}{k!}$$

In Exercise 30(a) we show that

$$\sum_{k=0}^{\infty} k \frac{\lambda^k}{k!} = \lambda e^{\lambda}$$

and therefore

$$E(X) = e^{-\lambda} \sum_{k=0}^{\infty} k \frac{\lambda^k}{k!} = e^{-\lambda} \lambda e^{\lambda} = \lambda$$

From

$$E(X^2) = \sum_{k=0}^{\infty} k^2 P(X = k) = \sum_{k=0}^{\infty} k^2 e^{-\lambda} \frac{\lambda^k}{k!} = e^{-\lambda} \sum_{k=0}^{\infty} k^2 \frac{\lambda^k}{k!}$$

using the formula for the infinite sum

$$\sum_{k=0}^{\infty} k^2 \frac{\lambda^k}{k!} = \lambda e^{\lambda} + \lambda^2 e^{\lambda}$$

that we derive in Exercise 30(b), we get

$$E(X^2) = e^{-\lambda} \left(\lambda e^{\lambda} + \lambda^2 e^{\lambda} \right) = \lambda + \lambda^2$$

The variance of X is

$$\text{var}(X) = E(X^2) - (E(X))^2 = \lambda + \lambda^2 - \lambda^2 = \lambda$$

Note that $E(X) = \text{var}(X) = \lambda$. The standard deviation of X is $\sqrt{\lambda}$.

If the mean number of occurrences λ is large, so is the standard deviation $\sqrt{\lambda}$. This means that the distribution is quite spread out (and thus not of much value). For this reason a Poisson distribution is usually used (and gives meaningful results, as we will see) when the mean number of occurrences is small, i.e., for rarely occurring random phenomena.

Example 12.3   Radiotherapy in Cancer Treatment

Radiotherapy, or radiation therapy, is a medical procedure that is used for a number of purposes, from cancer treatment, control of malignant cells, and palliative treatments to treatments of non-malignant conditions.

In the case of cancer, ionizing radiation (a beam of charged particles) is used to disable the development of cancerous cells by damaging their DNA, thus eventually killing them. The amount of radiation that is applied is of crucial importance in the treatment.

Assume that a cancer has N cells (a cancer starts with one cell, and can grow to contain more than one billion cells). As well, assume that one charged particle in the radiation beam that is used is capable of damaging one cell. We will say that a particle "hits" the cell (thus, one or more hits into the same cell will disable it). Finally, assume that the number of hits per cell is Poisson-distributed.

(a) Suppose that the radiation beam contains N charged particles (that's one particle per cancerous cell). How many cancerous cells will be missed in the treatment? What ratio of cells will be hit more than once?

(b) Suppose that the radiation beam contains $3N$ particles (that's three particles per cancerous cell). What ratio of cancerous cells will survive this treatment (i.e., will not be hit by a single particle)?

▶ Let X denote the number of particles that hit a cell.

(a) It is assumed that the average number of particles that hit a cell is 1. Since the expected value of X is λ, it follows that $\lambda = 1$; thus, $X \sim \text{Po}(1)$. The probability that a cell does not get hit is

$$P(X = 0) = e^{-1} \frac{1^0}{0!} = e^{-1} \approx 0.367879$$

Thus, over one third of all cancerous cells will be unaffected by this treatment (i.e., will not be hit by radiation). In other words, the treatment succeeds in killing about $(1 - 0.367879 = 0.632121)$ 63.2% of cells.

For practice, we compute

$$P(X = 1) = e^{-1} \frac{1^1}{1!} = e^{-1} \approx 0.367879$$

$$P(X = 2) = e^{-1} \frac{1^2}{2!} = e^{-1} \frac{1}{2} \approx 0.183940$$

$$P(X = 3) = e^{-1} \frac{1^3}{3!} = e^{-1} \frac{1}{6} \approx 0.061313$$

$$P(X = 4) = e^{-1} \frac{1^4}{4!} = e^{-1} \frac{1}{24} \approx 0.015328$$

and so on; see Figure 12.1 for the histogram. We see that about 37% of the cells will be hit once, about 18% will be hit twice, and so on.

To find out what ratio of cells will be hit more that once we compute

$$P(X > 1) = 1 - P(X \leq 1)$$
$$= 1 - P(X = 0) - P(X = 1)$$
$$\approx 1 - 0.367879 - 0.367879 = 0.264242$$

Thus, 26.4% of all cells will be hit more than once.

(b) This time, $X \sim \text{Po}(3)$, and

$$P(X = 0) = e^{-3} \frac{3^0}{0!} = e^{-3} \approx 0.0497871$$

So, about 5% of the cells will not be hit. The good news is that tripling the intensity of the radiation improves the ratio of cancerous cells killed from 63.2% to 95%. The bad news, though, is that 5% of cancerous cells will still survive the treatment.

Example 12.4   **Using the Poisson Distribution to Identify a Possible Epidemic Outbreak**

In a non-epidemic situation, occurrences of certain infections (or diseases, or medical conditions) are viewed as independent, random events and are often modelled with a Poisson distribution. If the actual number of cases deviates significantly from what the Poisson model predicts, there is a possibility of an outbreak. Let's look at an example.

According to the Meningitis Research Foundation of Canada, the number of occurrences of pneumococcal meningitis is 2 per 100,000 Canadians per year. [Source: www.meningitis.ca/en/what_is_meningitis/pneumococcal.shtml.] That is two cases per year in a city the size of St. John's (Newfoundland and Labrador) or Lethbridge (Alberta).

Assume that in the last month at least two cases of pneumococcal meningitis were diagnosed in St. John's. Is this normal (i.e., could it be attributed to the randomness of the occurrence of meningitis), or is there a reason for concern as this could be a sign of an epidemic outbreak?

▶ Two cases per year translate to $2/12 = 1/6$ cases per month. Thus, we model the number of cases X using the distribution $X \sim \text{Po}(1/6) \approx \text{Po}(0.166667)$. The probability that there are fewer than two cases in a month in St. John's is

$$P(X < 2) = P(X = 0) + P(X = 1)$$
$$= e^{-0.166667} \frac{0.166667^0}{0!} + e^{-0.166667} \frac{0.166667^1}{1!}$$
$$\approx 0.846481 + 0.141081 = 0.987562$$

In words, fewer than two cases of meningitis a month occur with 98.7% chance. So identifying two or more cases in a month is very unlikely to happen by sheer randomness (the chance of that happening is 1.3%). Thus, there are reasons to believe that there is an epidemic outbreak.

Example 12.5   **Chocolate with Almonds**

Divide a bar of chocolate into 24 squares of equal size (we call them squares, but they are not really squares), and count the number of pieces of almond in each square. In Figure 12.2 we simulated this situation by asking a computer to randomly select 72 points within the 6 by 4 grid (i.e., 24 squares). Thus, on average, there should be three pieces of almond in each square.

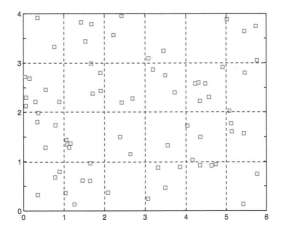

FIGURE 12.2

Simulation of the distribution of almond pieces in a bar of chocolate

In Table 12.1 we recorded the actual count (number of squares that contain a certain number of pieces of almond) and compared with the prediction given by $X \sim \text{Po}(3)$. In order to obtain a closer match, we would need to repeat the experiment many times and calculate the averages for the relative frequencies.

Table 12.1

Number of pieces of almond	Number of squares	Relative frequency	Poisson distribution
0	1	0.0416667	0.049787
1	2	0.0833333	0.149361
2	7	0.291667	0.224042
3	6	0.25	0.224042
4	5	0.208333	0.168031
5	1	0.0416667	0.100819
6	1	0.0416667	0.0504094
7	1	0.0416667	0.0216040

To calculate the probabilities in the last column, we used

$$P(X = k) = e^{-3} \frac{3^k}{k!}$$

with $k = 0, 1, 2, \ldots, 7$.

We discuss two more properties of the Poisson distribution.

Theorem 12   Sum of Poisson Distributions

Assume that X_1 and X_2 are independent, Poisson-distributed random variables, $X_1 \sim \text{Po}(\lambda_1)$ and $X_2 \sim \text{Po}(\lambda_2)$. Then their sum $X = X_1 + X_2$ is Poisson-distributed with $\lambda = \lambda_1 + \lambda_2$.

By the additive property of the expected value,

$$E(X) = E(X_1 + X_2) = E(X_1) + E(X_2) = \lambda_1 + \lambda_2$$

Since the two random variables are independent,

$$\text{var}(X) = \text{var}(X_1 + X_2) = \text{var}(X_1) + \text{var}(X_2) = \lambda_1 + \lambda_2$$

This, of course, does not prove that X is a Poisson random variable (but it's a good start; since the mean and the variance of X are equal, X does have a chance of being a Poisson distribution). To prove the theorem, we have to show that

$$P(X = k) = e^{-(\lambda_1 + \lambda_2)} \frac{(\lambda_1 + \lambda_2)^k}{k!}$$

for all $k = 0, 1, 2, \ldots$. We will show that the first two cases ($k = 0$ and $k = 1$) work, i.e.,

$$P(X = 0) = e^{-(\lambda_1 + \lambda_2)} \frac{(\lambda_1 + \lambda_2)^0}{0!} = e^{-(\lambda_1 + \lambda_2)}$$

$$P(X = 1) = e^{-(\lambda_1 + \lambda_2)} \frac{(\lambda_1 + \lambda_2)^1}{1!} = e^{-(\lambda_1 + \lambda_2)}(\lambda_1 + \lambda_2)$$

Using the assumption of independence,

$$
\begin{aligned}
P(X = 0) &= P(X_1 = 0 \text{ and } X_2 = 0) \\
&= P(X_1 = 0)P(X_2 = 0) \\
&= e^{-\lambda_1} \frac{\lambda_1^0}{0!} e^{-\lambda_2} \frac{\lambda_2^0}{0!} \\
&= e^{-\lambda_1} e^{-\lambda_2} = e^{-(\lambda_1 + \lambda_2)}
\end{aligned}
$$

Starting with mutual exclusivity and then using independence,

$$
\begin{aligned}
P(X = 1) &= P((X_1 = 0 \text{ and } X_2 = 1) \text{ or } (X_1 = 1 \text{ and } X_2 = 0)) \\
&= P(X_1 = 0 \text{ and } X_2 = 1) + P(X_1 = 1 \text{ and } X_2 = 0) \\
&= P(X_1 = 0)P(X_2 = 1) + P(X_1 = 1)P(X_2 = 0) \\
&= e^{-\lambda_1} \frac{\lambda_1^0}{0!} e^{-\lambda_2} \frac{\lambda_2^1}{1!} + e^{-\lambda_1} \frac{\lambda_1^1}{1!} e^{-\lambda_2} \frac{\lambda_2^0}{0!} \\
&= e^{-\lambda_1} e^{-\lambda_2} \lambda_2 + e^{-\lambda_1} \lambda_1 e^{-\lambda_2} \\
&= e^{-(\lambda_1 + \lambda_2)}(\lambda_1 + \lambda_2)
\end{aligned}
$$

The remaining cases use the same ideas, but are technically more involved.

Example 12.6   **Bacteria on a Kitchen Countertop**

A stainless steel countertop in a restaurant kitchen with a total area of 10 m^2 is known to contain 250 type A bacteria and 870 type B bacteria. A 100-cm^2 patch of the countertop is randomly chosen. What is the probability that there are no bacteria at all on it?

▶ The average number of type A bacteria is 250 per 10 m^2 = 100,000 cm^2 or 0.25 per 100 cm^2. Thus, we model the number of type A bacteria in a 100-cm^2 patch by a Poisson distribution $A \sim \text{Po}(0.25)$. The average number of type B bacteria is 870 per 100,000 cm^2 or 0.87 per 100 cm^2. Thus, the number of type B bacteria in a 100-cm^2 patch is Poisson distributed, $B \sim \text{Po}(0.87)$. We are asked to find $P(A + B = 0)$.

By Theorem 12, the random variable $A + B$ is Poisson-distributed with parameter $\lambda = 0.25 + 0.87 = 1.12$. We compute

$$P(A + B = 0) = e^{-1.12} \frac{1.12^0}{0!} = e^{-1.12} \approx 0.326280$$

Thus, there is about a 33% chance that a randomly selected 100-cm^2 patch of the countertop is free of bacteria.

The Poisson Distribution Approximates the Binomial Distribution

In Example 12.2 we modelled the number of emergency scanning requests during a 9-hour interval using the Poisson distribution $X \sim \text{Po}(6.56)$.

On average, there are $6.56/9$ requests per hour, or
$$6.56/(9 \cdot 3{,}600) = 0.000202469$$

requests per second. Assuming that the consecutive emergency scanning requests come more than 1 second apart, we can reformulate the situation in the following way. Define the Bernoulli experiment B_i by

$$B_i = \begin{cases} 1 & \text{emergency request received (success)} \\ 0 & \text{no emergency request} \end{cases}$$

for $i = 1, 2, \ldots, 9 \cdot 3{,}600 = 32{,}400$. The probability distribution for each B_i is given in Table 12.2.

Table 12.2

k	$P(B = k)$
0	0.999797531
1	0.000202469

Define by B the binomial variable that counts the number of successes in a 9-hour interval. Let's compare the two distributions, but using both to compute the probability that there are exactly eight emergency requests in a 9-hour interval. Using the Poisson distribution $X \sim \text{Po}(6.56)$, we get

$$P(X = 8) = e^{-6.56} \frac{6.56^8}{8!} \approx 0.120431$$

Using the binomial distribution (8 successes in 32,400 trials) we get

$$b(8, 32{,}400; 0.000202469) = \binom{32{,}400}{8} 0.000202469^8 (1 - 0.000202469)^{32{,}400-8}$$
$$\approx 0.122285$$

In Table 12.3 we compare a few more values.

Table 12.3

k	Poisson $X \sim \text{Po}(6.56)$ $P(X = k)$	Binomial $b(k, 32{,}400; 0.000202469)$
0	0.00141589	0.00143662
5	0.143340	0.145540
10	0.0575846	0.0584642
15	0.00194128	0.00196927
20	0.0000126760	0.0000128379

We divided the 9-hour interval into seconds. If we had we divided it into tenths of a second, we would have obtained an even closer approximation. Thus, we can view the Poisson distribution as the limit of infinitely many Bernoulli trials, each with an infinitesimal probability of success.

A commonly accepted rule states that if the probability of success p is small (smaller than 0.01) and the number of trials is large (larger than 100), then the binomial distribution can be closely approximated by the Poisson distribution. In particular,

$$b(k, n; p) \approx P(X = k)$$

where $X \sim \text{Po}(np)$.

Let B be the binomial distribution with parameters n and p and X the Poisson distribution with parameter $\lambda = np$. The two distributions have the same means $E(B) = E(X) = np$. Their variances differ: $\text{var}(B) = np(1-p)$, whereas $\text{var}(X) = np$. However, when p is small, then $1 - p \approx 1$, and thus $\text{var}(B) \approx \text{var}(X)$.

The two distributions differ in their ranges: the binomial distribution has non-zero probabilities for $k = 0, 1, 2, \ldots, n$, whereas X has non-zero probabilities for all $k = 0, 1, 2, \ldots$. For large k, though, $P(X = k)$ is very small.

Summary   The **Poisson process** describes events that are independent and occur at a constant rate λ. The associated **Poisson distribution** counts the number of events that occur in a given time or space interval. The sum of independent, Poisson-distributed random variables is Poisson-distributed. In certain situations, the Poisson distribution is a close approximation of the binomial distribution.

12   Exercises

1. Compute the values $P(X = k)$ for $k = 0, 1, 2, 3$, and 4 for a Poisson distribution with $\lambda = 2.5$.

2. Compute the values $P(X = k)$ for $k = 0, 1, 2, 3, 4, 5$, and 6 for a Poisson distribution with $\lambda = 4$.

3. Suppose that X is Poisson-distributed with mean $\lambda = 12$. Find $P(4 \leq X \leq 7)$.

4. Suppose that X is Poisson-distributed with mean $\lambda = 2.6$. Find $P(X < 3)$ and $P(1 \leq X \leq 3)$.

5. Let $X \sim \text{Po}(4)$. Find the probability that X is at most 3.

6. Let $X \sim \text{Po}(4)$. Find the probability that X is at least 3.

7–8 ▪ Find the value of the parameter λ for each Poisson distribution, knowing that it is an integer. Explain your reasoning.

7.

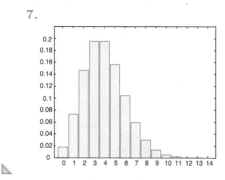

8.

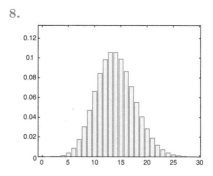

9. Certain types of a rare strain of respiratory infection occur in about 3 out of 2,000 people. During a particularly bad flu season, 12 out of 5,000 people were diagnosed with the infection. What is the probability of this event occurring?

10. The average number of more serious traffic accidents per week on the stretch of Highway 401 between Pearson International Airport and Yonge Street (downtown Toronto) is four. What is the probability that two accidents happen on the same day on that stretch of the highway?

11. The average number of more serious traffic accidents per week on the stretch of Highway 401 between Pearson International Airport and Yonge Street (downtown Toronto) is four. What is the probability that at least two accidents happen in a week on that stretch of the highway?

12. In a bag containing 30 apples, 2 are found to be spoiled. You buy a bag with 15 apples. What is the probability that you will find exactly 1 spoiled apple in the bag?

13. In a bag containing 30 apples, 2 are found to be spoiled. You buy a bag with 15 apples. What is the probability that you will find no more than two spoiled apples in the bag?

14. It has been determined that the average number of bacteria in a hot dog is four per 10 g. What is the probability that a 150-g hot dog will contain fewer than three bacteria?

15. A sample of 1 L of tap water has been found to contain six heavy metal particles. You drink a half-litre glass of tap water. How likely is it that you have not consumed any heavy metal particles?

16. A room is found to have seven dust particles per cubic centimetre of air. Find the probability that there are fewer than four dust particles in 1 cm^3 of air in the room.

17. Diffusing molecules leave a certain region at a rate of 0.4 molecules per hour. What is the probability that three or fewer will leave by the end of the second hour?

18. Diffusing molecules leave a certain region at a rate of 0.4 molecules per hour. What is the probability that more than two molecules will leave by the end of the third hour?

19. The rate at which we are hit by cosmic rays is about one per day. What is the probability that we will be hit at least once during an eight-hour interval?

20. Flying in an airplane, we get hit by cosmic rays more often than when we are on the ground. Assume that the rate at which we are hit is one per 4 hours. What is the probability that we will be hit between 10 and 12 times during an 8-hour flight? [Although the effects of cosmic radiation are negligible for an airplane passenger, they could have serious effects on the International Space Station personnel, who spend months at a time in space.]

21. A student receives text messages at a rate of three per hour. What is the probability that she receives more than five messages in an hour?

22. A student receives text messages at a rate of three per hour. What is the probability that she receives at least three messages in 2 hours?

23. Given that $X \sim \mathrm{Po}\,(1)$ and $Y \sim \mathrm{Po}\,(9)$. Assuming that X and Y are independent, find $P(X + Y = 2)$ and $P(Y = 2 \mid X + Y = 2)$.

24. Given that $X \sim \mathrm{Po}\,(5)$ and $Y \sim \mathrm{Po}\,(3)$. Assuming that X and Y are independent, find $P(X + Y = 3)$ and $P(X = 1 \mid X + Y = 3)$.

25. A student receives text messages at a rate of four per hour and (independent of it) phone calls at a rate of two per hour. Each phone call and each text message interrupts the student's work. How likely is it that the student will experience no interruptions in 1 hour? How likely is it that the student will experience one interruption every 10 minutes?

26. A student receives text messages at a rate of three per hour and (independent of it) phone calls at a rate of one per hour. Each phone call and each text message interrupts the student's work. Find the probability that the student will experience at least three interruptions in 30 minutes. Find the probability that the student will experience at least one interruption in 5 minutes.

27. About 3 in 1,000 people experience serious side effects from an allergy medication. Find the probability that in a group of 200 people nobody experiences serious side effects in two ways: using the binomial distribution, and then using the Poisson approximation. Compare the results.

28. About 3.8 in 1,000 births in Canada are affected by fetal alcohol syndrome (FAS). Find the probability that in 500 births there will be exactly one case of FAS in two ways: using the binomial distribution, and then using the Poisson approximation. Compare the results.

29. About 2 in 1,000 people suffer serious consequences from lactose intolerance. Find the probability that in a group of 500 people one person experiences serious consequences in two ways: using the binomial distribution, and then using the Poisson approximation. Compare the results.

30. Assume that the algebraic operations suggested here are true when applied to infinite sums.

    (a) Start with

    $$1 + \lambda + \frac{\lambda^2}{2!} + \frac{\lambda^3}{3!} + \cdots = e^\lambda$$

    differentiate with respect to λ, and then multiply by λ to show that

    $$\lambda + 2\frac{\lambda^2}{2!} + 3\frac{\lambda^3}{3!} + \cdots = \lambda e^\lambda$$

    (b) Start with the formula you proved in (a), differentiate with respect to λ again, and then multiply by λ to show that

    $$\sum_{k=0}^{\infty} k^2 \frac{\lambda^k}{k!} = \lambda + 2^2 \frac{\lambda^2}{2!} + 3^2 \frac{\lambda^3}{3!} + \cdots = \lambda e^\lambda + \lambda^2 e^\lambda$$

13 Continuous Random Variables

Departing from studying random variables that take on a finite number of values, we now focus on random variables whose range contains a **continuum of values.** The word "continuum" describes the size of the set of real numbers or intervals of real numbers. For instance, the sets $[3, 4)$, $(-\infty, 0)$, and $(-\infty, \infty)$ contain a continuum of real numbers. Unlike positive numbers or integers, a continuum of numbers cannot be written in a sequence.

Probability Density Function

Recall that a random variable is a function from a sample space into a subset of real numbers.

Definition 35 Continuous Random Variable

A random variable that takes on a continuum of values is called a *continuous random variable.*

As with derivatives and integrals in calculus, our approach to studying continuous random variables will involve limits of smaller and smaller quantities. We start with an example.

Example 13.1 Distributions of Lengths of Boa Constrictors

The boa constrictor (boa, for short) is a large species of snake that can grow to anywhere between 1 m and 4 m in length. The lengths of 500 boas are recorded in Table 13.1; we show the frequencies as well as the relative frequencies for the six length ranges.

Table 13.1

Length range (m)	Frequency	Relative frequency
$[1, 1.5)$	20	0.04
$[1.5, 2)$	58	0.116
$[2, 2.5)$	122	0.244
$[2.5, 3)$	180	0.36
$[3, 3.5)$	86	0.172
$[3.5, 4)$	34	0.068

The histogram representing the probabilities (relative frequencies) is drawn in Figure 13.1.

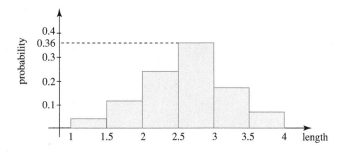

FIGURE 13.1

Histogram: the heights represent the probability

The horizontal axis represents the intervals (length ranges); we read the corresponding probabilities on the vertical axis. For instance, the probability that a randomly selected boa is between 2.5 m and 3 m in length is 0.36. Note that 0.36 is the *height of the rectangle* that represents the group of boas between 2.5 m and 3 m in length.

We now redraw the rectangles in the histogram so that their *areas, rather than their heights*, represent the probability; see Figure 13.2.

Consider the rectangle over $[2, 2.5)$. What should its height be? To satisfy

$$\text{area} = \text{base length} \cdot \text{height} = \text{probability}$$

we need

$$(2.5 - 2) \cdot \text{height} = 0.244$$

and thus the height must be $0.244/0.5 = 0.488$.

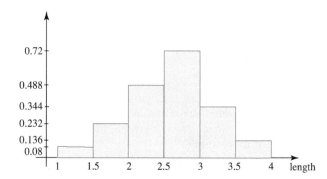

FIGURE 13.2

Histogram: the areas represent the probability

The height of the rectangle over $[3, 3.5)$ in Figure 13.2 is 0.344. The probability that a randomly chosen boa is between 3 m and 3.5 m long is equal to the area of the rectangle, $(0.5)(0.344) = 0.172$.

For various reasons, we might need to have a more precise probability mass function (note that, for instance, we put 3.04-m, 3.3-m and 3.48-m boas in the same box). Table 13.2 shows data arranged in twelve length ranges (rather than six), each of length 0.25 m.

Table 13.2

Length range (m)	Frequency	Relative frequency
$[1, 1.25)$	6	0.012
$[1.25, 1.5)$	14	0.028
$[1.5, 1.75)$	30	0.06
$[1.75, 2)$	28	0.056
$[2, 2.25)$	50	0.1
$[2.25, 2.5)$	72	0.144
$[2.5, 2.75)$	104	0.208
$[2.75, 3)$	76	0.152
$[3, 3.25)$	52	0.104
$[3.25, 3.5)$	34	0.068
$[3.5, 3.75)$	28	0.056
$[3.75, 4)$	6	0.012

Now we draw the histogram, with our new agreement that *areas* represent probabilities. This time, the length of each subinterval (length range) is 0.25, and so from $0.25 \cdot \text{height} = \text{probability}$ we get that

$$\text{height} = \text{probability}/0.25.$$

Thus, we multiply the probabilities (relative frequencies) in Table 13.2 by $1/0.25 = 4$ to get the heights of the rectangles in the histogram in Figure 13.3.

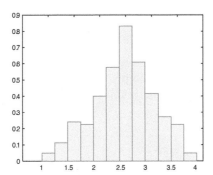

FIGURE 13.3

Histogram with shorter length ranges

The horizontal axis in Figure 13.3 represents the length ranges, but the vertical axis no longer represents the probabilities. As our calculation of the heights for the rectangles shows, the vertical axis represents probability/0.25, i.e., the probability per unit width of the length range.

Keeping the tradition that a quantity per unit length (or area, or volume) is called a density, we call the quantity on the vertical axis in Figure 13.3 the *probability density (function)*.

We continue increasing the number of subintervals by shrinking the length ranges on the horizontal axis. In Figure 13.4 we drew histograms based on 24 and 48 subintervals (the frequencies for the 48 subintervals are 0, 2, 1, 3, 3, 2, 5, 4, 7, 9, 5, 9, 5, 7, 7, 9, 9, 11, 12, 18, 14, 16, 20, 22, 20, 24, 25, 35, 30, 20, 16, 10, 19, 9, 13, 11, 11, 9, 9, 5, 9, 9, 6, 4, 0, 4, 1, 1; the frequencies for the 24 intervals are obtained by adding pairs of consecutive numbers in the list).

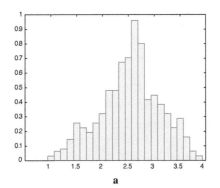 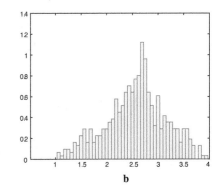

FIGURE 13.4

Histograms based on 24 and 48 subintervals

The shrinking rectangles that we obtain as we keep increasing the number of subintervals define the *probability density function* (think of the definite integral and approximating Riemann sums).

Looking at Figure 13.4b: how do we find the probability that a randomly chosen boa is between 1.75 m and 2 m in length?

Since the areas of the rectangles represent the probabilities, we add up the areas of those rectangles whose bases belong to $[1.75, 2)$; see Figure 13.5.

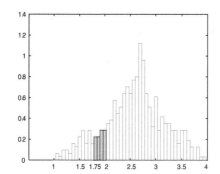

FIGURE 13.5

Probability of boa length between 1.75 m and 2 m

To summarize: by re-interpreting the histogram and by increasing the number of rectangles, we obtain the probability density function. The probability is the area of the region under the graph of the probability density function; see Figure 13.6.

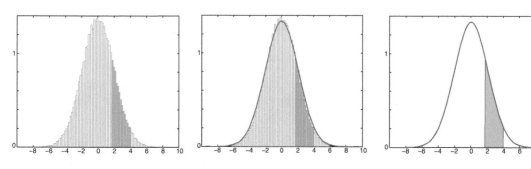

FIGURE 13.6

From a histogram to a density function

Before we learn how to work with probability density functions, we need to figure out which functions can possibly act as probability density functions.

Recall that the sum of the probabilities that define a distribution of a discrete random variable must add up to 1. Thus—in the limit—the area under the graph of a probability density function must be 1. The probability is a non-negative number, and so the probability density function must be non-negative.

In this, and in the forthcoming sections, we use the symbol I to denote any of the intervals $[a, b]$, (a, b), $(a, b]$, and $[a, b)$, including unbounded intervals such as $(-\infty, b)$, (a, ∞), and $(-\infty, \infty)$.

Definition 36 **Defining Properties of a Probability Density Function**

Assume that the interval I represents the sample space of an experiment (thus, a simple event is represented by a real number from I). A function $f(x)$ can be a probability density function if

(1) $f(x) \geq 0$ for all x in I.

(2) $\displaystyle\int_I f(x)dx = 1$.

Note that there is no requirement that $f(x) \leq 1$ on I, as was the case for the discrete probability distributions we studied earlier (see Example 13.2b).

If $I = [a, b]$, or (a, b), or $(a, b]$, or $[a, b)$, and a and b are real numbers, then $\int_I f(x)dx$ denotes the usual definite integral

$$\int_I f(x)\, dx = \int_a^b f(x)\, dx$$

If I is an unbounded interval, then $\int_I f(x)\,dx$ is an improper integral. For instance,

$$\int_a^\infty f(x)\,dx = \lim_{T\to\infty} \int_a^T f(x)\,dx$$

Example 13.2 **Probability Density Functions**

Show that

(a) $f(x) = x/2$ could be a probability density function on the interval $I = [0, 2]$.

(b) $f(x) = 3e^{-3x}$ could be a probability density function on $I = [0, \infty)$.

▶ (a) Clearly, $f(x) \geq 0$ for all x in $[0, 2]$. As well,

$$\int_0^2 \frac{x}{2}\,dx = \left.\frac{x^2}{4}\right|_0^2 = \frac{4}{4} - 0 = 1$$

(b) Because the exponential function is positive for all real numbers, it follows that $f(x) \geq 0$ for all $x \in [0, \infty)$. To check that the integral of f is 1, we need to use improper integration:

$$\int_0^\infty 3e^{-3x}\,dx = \lim_{T\to\infty} \int_0^T 3e^{-3x}\,dx$$

$$= \lim_{T\to\infty} \left.\left(-e^{-3x}\right)\right|_0^T$$

$$= \lim_{T\to\infty} \left(-e^{-3T} + e^0\right)$$

$$= -e^{-\infty} + 1 = 1$$

Note that $f(x) > 1$ for some values of x; see Figure 13.7.

FIGURE 13.7

The graph of $f(x) = 3e^{-3x}$

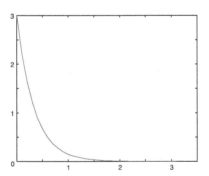

How do we calculate a probability using a probability density function?

Suppose that we conduct an experiment whose sample space is an interval I of real numbers, and whose outcomes are defined by a function $f(x)$ that satisfies properties (1) and (2) from Definition 36. The probability

$$P(a \leq X \leq b)$$

that an outcome X is between a and b is the area under the graph of $f(x)$ on $[a, b]$. Since the area is calculated as a definite integral, we get

$$P(a \leq X \leq b) = \int_a^b f(x)\,dx \qquad (13.1)$$

This formula holds for unbounded intervals as well, in which case the integral on the right side is an improper integral.

In the same sense as integrating the mass density gives the mass (of an object), integrating the probability density gives the probability (of an event occurring).

Using (13.1) with $a = b$, we get

$$P(a \leq X \leq a) = \int_a^a f(x)\,dx = 0$$

i.e.,

$$P(X = a) = 0 \qquad\qquad (13.2)$$

Thus, the probability that an outcome is *equal* to a particular real number is zero. In other words, what makes sense in the continuous case is to calculate the probability that the values of a random variable belong to an *interval* of real numbers. We will illustrate this point in numerous examples.

As a consequence of (13.2), we obtain

$$P(a \leq X \leq b) = P(a \leq X < b \text{ or } X = b)$$
$$= P(a \leq X < b) + P(X = b)$$
$$= P(a \leq X < b)$$

since $P(X = b) = 0$. Proceeding in the same way, we show that

$$P(a \leq X \leq b) = P(a \leq X < b) = P(a < X \leq b) = P(a < X < b)$$

Thus, including or excluding the endpoints of an interval does not affect the probability.

Example 13.3   Uniform Distribution

Consider the *uniform distribution* given by the probability density function $f(x) = 1$ on $[0, 1]$. Let X be a continuous random variable whose probability density function is $f(x)$. Find the probability that the value of X is between 0.17 and 0.28.

▶ The probability is given by

$$P(0.17 \leq X \leq 0.28) = \int_{0.17}^{0.28} 1\,dx = x\Big|_{0.17}^{0.28} = 0.28 - 0.17 = 0.11$$

According to the interpretation of the probability, the area of the shaded region below the graph of $f(x) = 1$ in Figure 13.8 is equal to 0.11.   ◢◣

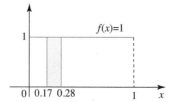

FIGURE 13.8

Interpreting the probability
$P(0.17 \leq X \leq 0.28)$ as area

Let's continue with the example: the probability that X is between 0.17 and 0.18 is

$$P(0.17 \leq X \leq 0.18) = \int_{0.17}^{0.18} 1\,dx = x\Big|_{0.17}^{0.18} = 0.18 - 0.17 = 0.01$$

Likewise, the probability that X is between 0.17 and 0.17001 is

$$P(0.17 \leq X \leq 0.17001) = \int_{0.17}^{0.17001} 1\,dx = 0.17001 - 0.17 = 0.00001$$

Although the probability that X is equal to a particular number is zero, the probability that X belongs to an interval, no matter how small, is not zero. In this case, the smaller the interval, the smaller the probability; so the fact that $P(X = 0.17) = 0$ is not surprising: if we keep shrinking the interval, the probability will approach zero.

The uniform distribution can be used to represent a random number generator, i.e., a piece of software that makes picking any number between 0 and 1 equally likely (in theory). Computers, no matter how powerful or sophisticated, work with *finite* numbers, so it makes sense to talk about the non-zero probability of picking a single number. The uniform distribution is an idealization in the case of (true) real numbers.

Example 13.4   Probability of a Virus Appearing in a Population

Let T be a random variable that tracks the time of the appearance of a virus within a population. At the start of the experiment ($t = 0$) the virus is not present, but it can appear at any time $t > 0$ (t is measured in days).

Unlike the situations we studied in previous sections, the random variable T can take on any non-negative real-number value (thus, it is a continuous random variable). The probability density function of T is given by

$$f(t) = 0.2e^{-0.2t}$$

for $t \geq 0$.

(a) Find the probability that the virus appears within the first 5 days.

(b) Find the probability that the virus appears on the sixth day.

(c) Find the probability that the virus appears between 6 a.m. and noon on the sixth day.

▶ The first day is defined by $0 \leq t \leq 1$, the second day by $1 \leq t \leq 2$, the fifth day by $4 \leq t \leq 5$, and so on. Recall that it makes no difference whether we use $\leq$ or $<$ (and thus we can, luckily, avoid discussing when exactly a day starts or ends).

(a) To find the probability, we integrate the probability density function:

$$P(0 < T \leq 5) = \int_0^5 0.2e^{-0.2t}\, dt$$
$$= 0.2\, \frac{1}{-0.2}\, e^{-0.2t}\Big|_0^5$$
$$= \left(-e^{-0.2t}\right)\Big|_0^5$$
$$= \left(-e^{-0.2(5)}\right) - \left(-e^{-0.2(0)}\right)$$
$$= -e^{-1} + 1 \approx 0.632$$

Note that we used the integration formula

$$\int e^{at}\, dt = \frac{1}{a}\, e^{at} + C,$$

which can be found by guessing and checked by differentiation.

Thus, there is about a 63.2% chance that the virus will appear within the first 5 days. This probability is equal to the area of the shaded region in Figure 13.9.

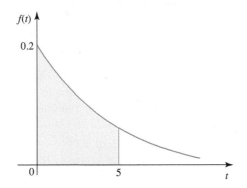

FIGURE 13.9

Geometric interpretation of the probability in Example 13.4(a)

(b) The probability that the virus appears on the sixth day is $P(5 \leq T \leq 6)$. Integrating, we get

$$
\begin{aligned}
P(5 \leq T \leq 6) &= \int_5^6 0.2e^{-0.2t}\, dt \\
&= \left(-e^{-0.2t}\right)\Big|_5^6 \\
&= -e^{-1.2} + e^{-1} \approx 0.067
\end{aligned}
$$

(c) The sixth day is defined by $5 \leq t \leq 6$, and so the probability that the virus appears between 6 a.m. and noon on the sixth day is

$$
\begin{aligned}
P(5.25 \leq T \leq 5.5) &= \int_{5.25}^{5.5} 0.2e^{-0.2t}\, dt \\
&= \left(-e^{-0.2t}\right)\Big|_{5.25}^{5.5} \\
&= -e^{-1.1} + e^{-1.05} \approx 0.017
\end{aligned}
$$

In theory, we can calculate the probability of the virus appearing during any time interval, no matter how small. In practice, the answers to (a) and (b) might be quite useful, whereas the answer to (c) might not mean much.

Cumulative Distribution Function

We can rephrase the answer to part (a) of Example 13.4 in the following way: the probability that the virus appears before $t = 5$ (i.e., by the end of day five) is

$$
P(T \leq 5) = \int_0^5 f(t)\, dt \approx 0.632
$$

In general, the probability that the virus appears before time x is

$$
P(T \leq x) = \int_0^x f(t)\, dt \tag{13.3}
$$

The function appearing on the right side in (13.3) is an *integral function*. Because it's important in probability, we give it a name.

Definition 37   Cumulative Distribution Function

Suppose that $f(x)$ is a probability density function defined on an interval $[a, b]$. The function $F(x)$ defined by

$$
F(x) = \int_a^x f(t)\, dt
$$

for all x in $[a, b]$ is called a *cumulative distribution function* of $f(x)$.

As mentioned earlier, the interval $[a, b]$ could be replaced by other intervals (such as (a, b) or $(a, b]$), or by unbounded intervals. For instance, the cumulative distribution function of a probability density function $f(x)$ defined on $(-\infty, \infty)$ is given by

$$
F(x) = \int_{-\infty}^x f(t)\, dt
$$

for all x in $(-\infty, \infty)$. Since we cannot use the same variable name for both the integrand and the upper limit, we renamed (as is common practice) the variable in the integrand to t.

Using the language of calculus, we say that the cumulative distribution function is the integral function of a probability density function. By convention, we use the same letter for both functions: uppercase for the cumulative distribution function and lowercase for the probability density function.

Example 13.5 Cumulative Distribution Function for Example 13.4

Consider the probability density function $f(t) = 0.2e^{-0.2t}$ defined on $[0, \infty)$. The associated cumulative distribution function is

$$F(x) = \int_0^x f(t)\, dt$$

$$= \int_0^x 0.2e^{-0.2t}\, dt$$

$$= \left(-e^{-0.2t}\right)\Big|_0^x$$

$$= \left(-e^{-0.2(x)}\right) - \left(-e^{-0.2(0)}\right)$$

$$= 1 - e^{-0.2x}$$

where $x \in [0, \infty)$. The probability that the virus appears by the end of day five (i.e., $x = 5$) is

$$P(T \le 5) = F(5) = 1 - e^{-0.2(5)} \approx 0.632.$$

Example 13.6 Cumulative Distribution Function of the Uniform Distribution

The uniform distribution is given by the probability density function $f(x) = 1$, $0 \le x \le 1$. The corresponding cumulative distribution function is

$$F(x) = \int_0^x 1\, dt = 1\Big|_0^x = x$$

where $0 \le x \le 1$. Let X be a continuous random variable uniformly distributed on $[0, 1]$ (in other words, $f(x) = 1$, $x \in [0, 1]$, is its probability density function). The probability that X is smaller than 0.76 is

$$P(X \le 0.76) = F(0.76) = 0.76$$

Thus, we can calculate the probability from either the probability density function or the cumulative distribution function. If we use the probability density function, the probability $P(X \le 0.76)$ is the area (Figure 13.10a). If we use the cumulative distribution function, $P(X \le 0.76)$ is the height, i.e., the value of F at 0.76. (Figure 13.10b).

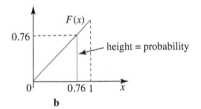

FIGURE 13.10

Geometric representations of the probability

Assume that f is a probability density function and F its associated cumulative distribution function:

$$F(x) = \int_a^x f(t)\, dt$$

(a could be a real number or $-\infty$). Using the Fundamental Theorem of Calculus we conclude that

(a) $F'(x) = f(x)$ (i.e., F is an antiderivative of f).

(b) For any numbers c and d in the domains of f and F,

$$\int_c^d f(t)\, dt = F(d) - F(c)$$

Statement (a) says that, given the cumulative distribution function, we find the probability density function by differentiating (by drawing the slopes of F). For instance, the derivative of $F(x) = 1 - e^{-0.2x}$ is $F'(x) = 0.2e^{-0.2x} = f(x)$ (Example 13.5); the derivative of $F(x) = x$ is $F'(x) = 1 = f(x)$ (Example 13.6).

Given a cumulative distribution function, we use (b) to find the probability:

$$P(c \leq x \leq d) = \int_c^d f(t)\, dt = F(d) - F(c)$$

So, in general, the probability is the *difference* of the values of the cumulative distribution function.

Example 13.7 Revisiting the Calculation of the Probability in Example 13.4

The cumulative distribution function for the appearance of the virus is given by $F(x) = 1 - e^{-0.2x}$ (we calculated it in Example 13.5).

The probability that the virus appears on the sixth day is

$$P(5 \leq T \leq 6) = F(6) - F(5)$$
$$= \left(1 - e^{-0.2(6)}\right) - \left(1 - e^{-0.2(5)}\right)$$
$$= -e^{-1.2} + e^{-1} \approx 0.067$$

confirming our answer to part (b) of Example 13.4.

We continue analyzing the cumulative distribution function

$$F(x) = \int_a^x f(t)\, dt$$

defined on an interval $[a, b]$.

Since $f(t) \geq 0$ for all t, it follows that $F(x) \geq 0$ for all x. As well (again using the fact that $f(t) \geq 0$),

$$F(x) = \int_a^x f(t)\, dt \leq \int_a^b f(t)\, dt = 1$$

by (2) in Definition 36.

Moreover, if a is a real number, then

$$F(a) = \int_a^a f(t)\, dt = 0$$

If $a = -\infty$, then

$$\lim_{x \to -\infty} F(x) = \lim_{x \to -\infty} \int_{-\infty}^x f(t)\, dt = 0$$

If b is a real number then

$$F(b) = \int_a^b f(t)\, dt = 1$$

Otherwise, if $b = \infty$, then

$$\lim_{x \to \infty} F(x) = \lim_{x \to \infty} \int_a^x f(t)\, dt = 1$$

by (2) in Definition 36.

From $F'(x) = f(x) \geq 0$, we conclude that $F(x)$ is a non-decreasing function. According to the Fundamental Theorem of Calculus, if f is continuous, so is F.

To summarize:

Theorem 13 **Properties of the Cumulative Distribution Function**

Assume that f is a probability density function, defined and continuous on an interval $[a, b]$. The left end a could be a real number or $-\infty$; the right end b could be a real number or ∞. Denote by F the associated cumulative distribution function. Then

(1) $0 \leq F(x) \leq 1$ for all x in $[a, b]$.

(2) $F(x)$ is continuous and non-decreasing.

(3) $\lim\limits_{x \to a} F(x) = 0$ and $\lim\limits_{x \to b} F(x) = 1$.

Note that f does not have to be continuous for F to be continuous (see Example 13.8). In all situations that we will encounter f will be continuous or will have a finite number of jump discontinuities (such as the function in Example 13.8).

Example 13.8 Finding the Cumulative Distribution Function

Consider the function

$$f(x) = \begin{cases} 0 & x < 0 \\ 1/2 & 0 \leq x \leq 2 \\ 0 & x > 2 \end{cases}$$

defined on $(-\infty, \infty)$.

(a) Show that f could be a probability density function.

(b) Find the corresponding cumulative distribution function.

▶ (a) Clearly, $f(x) \geq 0$ for all $x \in (-\infty, \infty)$ and

$$\int_{-\infty}^{\infty} f(x)\, dx = \int_{-\infty}^{0} f(x)\, dx + \int_{0}^{2} f(x)\, dx + \int_{2}^{\infty} f(x)\, dx$$

$$= \int_{0}^{2} \frac{1}{2}\, dx$$

$$= \frac{1}{2} x \Big|_{0}^{2} = 1$$

Thus, f satisfies both conditions of Definition 36.

(b) The cumulative distribution function is given by

$$F(x) = \int_{-\infty}^{x} f(t)\, dt$$

where $x \in (-\infty, \infty)$. If $x < 0$, then

$$F(x) = \int_{-\infty}^{x} f(t)\, dt = \int_{-\infty}^{x} 0\, dt = 0$$

If $0 \leq x \leq 2$, then

$$F(x) = \int_{-\infty}^{x} f(t)\, dt$$

$$= \int_{-\infty}^{0} f(t)\, dt + \int_{0}^{x} f(t)\, dt$$

$$= \int_{-\infty}^{0} 0\, dt + \int_{0}^{x} \frac{1}{2}\, dt = \frac{1}{2} x$$

If $x \geq 2$,

$$F(x) = \int_{-\infty}^{x} f(t)\,dt$$

$$= \int_{-\infty}^{0} f(t)\,dt + \int_{0}^{2} f(t)\,dt + \int_{2}^{x} f(t)\,dt$$

$$= \int_{-\infty}^{0} 0\,dt + \int_{0}^{2} \frac{1}{2}\,dt + \int_{2}^{x} 0\,dt$$

$$= 0 + 1 + 0 = 1$$

Thus, the cumulative distribution function is

$$F(x) = \begin{cases} 0 & x < 0 \\ x/2 & 0 \leq x \leq 2 \\ 1 & x > 2 \end{cases}$$

In Figure 13.11 we sketched both $f(x)$ and $F(x)$. Note that f is not continuous (at 0 and at 2), but, nevertheless, F is continuous for all x.

FIGURE 13.11

The functions $f(x)$ and $F(x)$ from Example 13.8

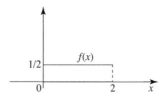
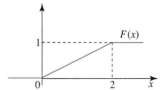

Example 13.9   Working with the Cumulative Distribution Function

Let

$$F(x) = \begin{cases} 0 & x < 1 \\ 1 - 1/x^2 & x \geq 1 \end{cases}$$

(a) Check that $F(x)$ satisfies properties (1), (2), and (3) from Theorem 13.

(b) Find the associated probability density function $f(x)$.

(c) Let X be a continuous random variable whose probability density function is $f(x)$. Find the probability $P(-2 \leq X \leq 4)$ in two different ways: by using the cumulative distribution function from (a) and by using the probability density function from (b).

▶ (a) Clearly, $F(x) \geq 0$ for all x. As well, $1 - 1/x^2 \leq 1$ for $x \geq 1$. Since $F(x) = 0$ for $x < 1$, we conclude that, for any x, $F(x) \leq 1$.

The function $F(x)$ is continuous for all $x \neq 1$. From

$$\lim_{x \to 1^+} F(x) = \lim_{x \to 1^+} \left(1 - \frac{1}{x^2}\right) = 0$$

$$\lim_{x \to 1^-} F(x) = \lim_{x \to 1^-} 0 = 0$$

and $F(1) = 0$ we conclude that F is continuous at $x = 1$. Thus, F is continuous at all real numbers x.

Calculating the derivative

$$F'(x) = \left(1 - \frac{1}{x^2}\right)' = \frac{2}{x^3} \geq 0$$

we conclude that $F(x)$ is increasing if $x \geq 1$. Because $F(x) = 0$ if $x < 1$, it follows that F is non-decreasing.

Finally,

$$\lim_{x \to -\infty} F(x) = \lim_{x \to -\infty} 0 = 0$$

and

$$\lim_{x \to \infty} F(x) = \lim_{x \to \infty} \left(1 - \frac{1}{x^2}\right) = 1 - 0 = 1$$

and we are done: all of (1), (2), and (3) from Theorem 13 are satisfied.

The graph of $F(x)$ is shown in Figure 13.12.

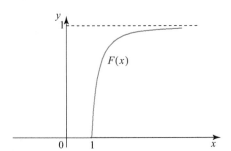

FIGURE 13.12

The cumulative distribution function $F(x)$

(b) Recall that $f(x) = F'(x)$. Thus,

$$f(x) = \begin{cases} 0 & x < 1 \\ 2/x^3 & x > 1 \end{cases}$$

Note that F is not differentiable at $x = 1$ (it has a corner there) and so $f(1)$ is not defined. As is common in these situations, to make f defined for all x, we set $f(1) = 0$. Thus,

$$f(x) = \begin{cases} 0 & x \le 1 \\ 2/x^3 & x > 1 \end{cases}$$

(c) Using the cumulative distribution function,

$$P(-2 \le X \le 4) = F(4) - F(-2) = \left(1 - \frac{1}{4^2}\right) - 0 = \frac{15}{16}$$

Using the probability density function,

$$P(-2 \le X \le 4) = \int_{-2}^{4} f(t)\, dt$$

$$= \int_{-2}^{1} f(t)\, dt + \int_{1}^{4} f(t)\, dt$$

$$= \int_{-2}^{1} 0\, dt + \int_{1}^{4} \frac{2}{t^3}\, dt$$

$$= 0 - \frac{1}{t^2}\Big|_{1}^{4}$$

$$= -\frac{1}{4^2} - (-1) = \frac{15}{16}$$

The next example is a bit more challenging.

Example 13.10 Calculating the Cumulative Distribution Function

Let $f(x) = xe^{-x}$, where $x \in [0, \infty)$. Check that f could be a probability density function, and find the corresponding cumulative distribution function.

▶ Clearly, $f(x) \ge 0$ for all x in $[0, \infty)$. To integrate $f(x)$, we use integration by parts.

Let $u = x$ and $v' = e^{-x}$. Then $u' = 1$, $v = -e^{-x}$, and

$$\int xe^{-x}\, dx = uv - \int vu'\, dx$$

$$= -xe^{-x} + \int e^{-x}\, dx$$

$$= -xe^{-x} - e^{-x} + C$$

Thus

$$\int_0^\infty xe^{-x}\, dx = \lim_{T \to \infty} \int_0^T xe^{-x}\, dx$$

$$= \lim_{T \to \infty} \left(-xe^{-x} - e^{-x}\right)\Big|_0^T$$

$$= \lim_{T \to \infty} \left(-Te^{-T} - e^{-T}\right) - (0 - 1) = 1$$

since

$$\lim_{T \to \infty} e^{-T} = 0$$

and, by L'Hôpital's rule,

$$\lim_{T \to \infty} Te^{-T} = \lim_{T \to \infty} \frac{T}{e^T} = \lim_{T \to \infty} \frac{1}{e^T} = 0$$

Thus, f satisfies both conditions of Definition 36. The cumulative distribution function is given by

$$F(x) = \int_0^x f(t)\, dt = \int_0^x te^{-t}\, dt$$

$$= \left(-te^{-t} - e^{-t}\right)\Big|_0^x$$

$$= \left(-xe^{-x} - e^{-x}\right) - (0 - 1) = 1 - xe^{-x} - e^{-x}$$

The graphs of f and F are shown in Figure 13.13.

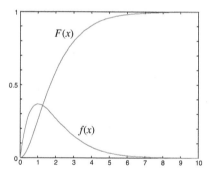

FIGURE 13.13

The graphs of f and F from Example 13.10

The Mean and the Variance

Analogous to the discrete case, we define the mean and the variance for continuous random variables.

Definition 38   The Mean and the Variance of a Continuous Random Variable

Let X be a continuous random variable with probability density function $f(x)$, defined on an interval $[a, b]$. As usual, a could be a real number or $-\infty$; b could be a real number or ∞. The *mean* (or the *expected value*) of X is given by

$$\mu = E(X) = \int_a^b xf(x)\, dx$$

The expected value of $g(X)$ (i.e., the expected value of a function of X) is

$$E(g(X)) = \int_a^b g(x)f(x)\,dx$$

The *variance* of X is

$$\text{var}(X) = E\left[(X - \mu)^2\right] = \int_a^b (x - \mu)^2 f(x)\,dx$$

The variance can also be calculated from

$$\text{var}(X) = E(X^2) - (E(X))^2 = \int_a^b x^2 f(x)\,dx - \left(\int_a^b x f(x)\,dx\right)^2$$

Example 13.11 The Mean and the Variance of the Uniform Distribution

Find the mean and the variance of the random variable X distributed uniformly on $[0, 1]$.

▶ Recall that the probability density function of X is given by $f(x) = 1$, $x \in [0, 1]$. The mean is, not surprisingly,

$$\mu = \int_0^1 x f(x)\,dx = \int_0^1 x\,dx = \frac{x^2}{2}\Big|_0^1 = \frac{1}{2}$$

The variance is

$$\begin{aligned}
\text{var}(X) &= \int_0^1 (x - \mu)^2 f(x)\,dx \\
&= \int_0^1 \left(x - \frac{1}{2}\right)^2 dx \\
&= \frac{1}{3}\left(x - \frac{1}{2}\right)^3 \Big|_0^1 \\
&= \frac{1}{3}\left(\frac{1}{2}\right)^3 + \frac{1}{3}\left(\frac{1}{2}\right)^3 = \frac{1}{12}
\end{aligned}$$

Example 13.12 Calculating the Mean and the Variance

The probability density function of a continuous random variable X is given by

$$f(x) = \begin{cases} 4x^3 & 0 \le x \le 1 \\ 0 & \text{otherwise} \end{cases}$$

Find its mean and variance.

▶ The mean of X is

$$\begin{aligned}
\mu = E(X) &= \int_{-\infty}^{\infty} x f(x)\,dx \\
&= \int_{-\infty}^{0} x f(x)\,dx + \int_0^1 x f(x)\,dx + \int_1^{\infty} x f(x)\,dx \\
&= \int_{-\infty}^{0} 0\,dx + \int_0^1 x(4x^3)\,dx + \int_1^{\infty} 0\,dx \\
&= 4\frac{x^5}{5}\Big|_0^1 = \frac{4}{5}
\end{aligned}$$

Note that all we have to do is to integrate over the interval $[0, 1]$ where $f(x)$ is not zero. To calculate the variance, we use $\text{var}(X) = E(X^2) - (E(X))^2$. From

$$E(X^2) = \int_{-\infty}^{\infty} x^2 f(x)\,dx = \int_0^1 x^2(4x^3)\,dx = 4\frac{x^6}{6}\Big|_0^1 = \frac{2}{3}$$

we obtain

$$\text{var}(X) = E(X^2) - (E(X))^2 = \frac{2}{3} - \left(\frac{4}{5}\right)^2 = \frac{2}{75}$$

Remark   In a number of statements in this section, we have used the phrase "probability density function $f(x)$, defined on an interval $[a, b]$; a could be a real number or $-\infty$; b could be a real number or ∞." All we wanted to say is that any type of interval (bounded or not) is allowed as the domain of $f(x)$.

Some textbooks use the following approach, which is only superficially different. Assume that a probability density function is defined on some interval $[a, b]$. We can extend its domain to $(-\infty, \infty)$ by defining $f(x)$ to be 0 outside $[a, b]$ (for an illustration, see Example 13.12). So one can *assume* from the start that the domain of a probability density function is $(-\infty, \infty)$.

Having defined the probability density function $f(x)$ on $(-\infty, \infty)$, we define all quantities related to it using integration from $-\infty$ to ∞. For instance,

$$\mu = \int_{-\infty}^{\infty} x f(x)\, dx$$

or

$$\text{var}(X) = \int_{-\infty}^{\infty} (x - \mu)^2 f(x)\, dx$$

and so on; see Examples 13.8 and 13.12 to convince yourself that this approach actually does not differ at all from the way we have done it.

Small point: extending a probability density function beyond its domain to $(-\infty, \infty)$, although mathematically sound, might not make sense in the context of an application. In Example 13.13 the variable x represents distance, so using negative numbers does not make sense.

Example 13.13   Mean Dispersal Distance

In studying certain aspects of the development of an ecosystem, researchers use probability to model the dispersal of plant seeds by various species, such as bats, birds, and lizards.

The function $f(x) = ae^{-ax}$, where $a > 0$ and $x \in [0, \infty)$, is used as the density function, in the sense that the integral

$$r(c) = \int_{0}^{c} f(x)\, dx = \int_{0}^{c} ae^{-ax}\, dx$$

gives the ratio of seeds dispersed within the circular region of radius c centred at the source of the seeds, taken to be at $x = 0$. (This circular region is sometimes referred to as the "seed shadow.") Note that

$$\int_{0}^{\infty} ae^{-ax}\, dx = 1$$

(see Exercise 34(a)), so the ratio $r(c)$ is a number between 0 and 1.

What is the mean seed dispersal distance?

▶ Note that the model allows for arbitrarily large distances of seed dispersal ($x \in [0, \infty)$), which is of course not realistic. However, the model works because the exponential function decays quickly (for instance, $e^{-20} \approx 2 \cdot 10^{-9}$, $e^{-50} \approx 1.9 \cdot 10^{-22}$, and $e^{-100} \approx 3.7 \cdot 10^{-44}$). Thus, we can safely neglect the effects of distances beyond a certain (relatively small) range.

The average dispersal distance is

$$\int_{0}^{\infty} x f(x)\, dx = \int_{0}^{\infty} axe^{-ax}\, dx$$

Using integration by parts (as in Example 13.10; see Exercise 34(b) for details) we get

$$\int axe^{-ax}\, dx = -\frac{1}{a}e^{-ax} - xe^{-ax} + C$$

Thus,

$$\int_0^\infty axe^{-ax}\, dx = \left(-\frac{1}{a}e^{-ax} - xe^{-ax} \right)\Bigg|_0^\infty = 0 - \left(\frac{1}{a} \right) = \frac{1}{a}$$

(see Exercise 34(c) for the details of the calculation of the improper integral). The mean seed dispersal distance is the reciprocal of the constant a (which needs to be determined experimentally).

The Median

In discussing the median for finite random variables, we mentioned that it makes the most sense when the random variable involved takes on a large number of different values. Continuous random variables certainly possess this property.

Recall that the median is the value that divides the sample into two equally likely events. In other words, the median is the point where the cumulative distribution function is equal to $1/2$. (That the median exists is a consequence of the Intermediate Value Theorem; see Exercise 35.)

In Figure 13.14 we show how to identify the median from both a probability density function and a cumulative distribution function.

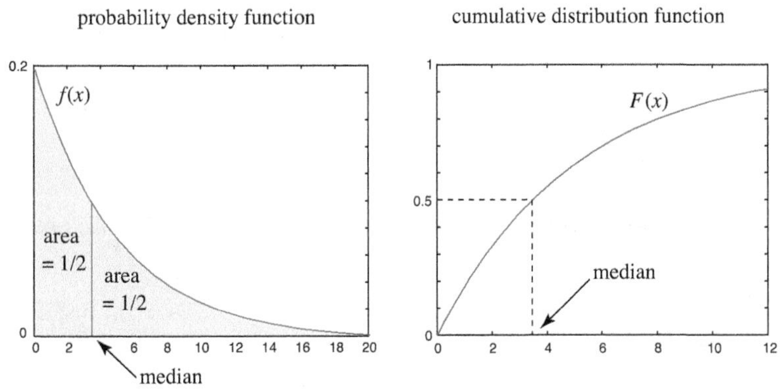

FIGURE 13.14

Locating the median

Example 13.14   **The Median of the Distribution in Example 13.5**

The cumulative distribution function related to the appearance of a virus in a population is given by $F(x) = 1 - e^{-0.2x}$. Find the median time of the appearance of the virus.

▶ The median is the value of x where $F(x) = 1 - e^{-0.2x} = 0.5$. Solving for x, we get

$$e^{-0.2x} = 0.5$$
$$-0.2x = \ln 0.5$$
$$x = \frac{\ln 0.5}{-0.2} \approx 3.47$$

Thus, the median time is about 3.5 days.

Summary   **Continuous random variables** take on a continuum of values. By modifying the way we draw histograms so that the areas, rather than the heights, represent the probability, we constructed the **probability density function** of a continuous random variable. To find the probability, we integrate the probability density function. The **cumulative distribution function** is defined in the same way as for discrete random variables. Replacing the sum by the definite integral in the definitions for the discrete variables, we obtain the **mean** and the **variance** for continuous random variables.

13   Exercises

1. Explain why $f(x) = 1 - x^2$, $x \in [0, 2]$, cannot be a probability density function of any random variable.

2. Explain why $f(x) = 1/2$, $x \in [1, 2]$, cannot be a probability density function of any random variable.

3. Find the value of the constant a so that $f(x) = a/x$, $1 \leq x \leq 10$, satisfies the properties of the probability density function.

4. Find the value of the constant a so that $f(x) = ax(1 - x)$, $0 \leq x \leq 1$, satisfies the properties of the probability density function.

5. Find the value of the constant a so that $f(x) = a(1 + x^2)^{-1}$, $x \in (0, \infty)$, satisfies the properties of the probability density function.

6. Check that the function $f(x) = \dfrac{1}{2\sqrt{x}}$, $0 < x \leq 1$, can be a probability density function. Find its mean.

7. Check that the function $f(x) = \dfrac{2}{x^3}$, $x \in [1, \infty)$, can be a probability density function. Find its mean.

8. Consider uniformly distributed random variables X_1 and X_2 whose probability density functions are $f_1(x) = 1/2$, $0 \leq x \leq 2$, and $f_2(x) = 1/10$, $0 \leq x \leq 10$. Which one has the larger variance?

9. The uniform distribution is characterized by the fact that its probability density function is a constant function. Can there be a uniform distribution on $[0, \infty)$?

10. Find the value of c so that $f(x) = c$, $a \leq x \leq b$, is a probability density function (a and b are real numbers). The random variable X whose probability density function is $f(x)$ is said to be *distributed uniformly* on $[a, b]$. Can you guess what the mean of X is? Find the mean and the variance of X.

▽ 11–14 ▪ Find each probability in two ways: using the probability density function $f(x)$ of the random variable X and using the corresponding cumulative distribution function.

11. $f(x) = 0.3 + 0.2x$, $0 \leq x \leq 2$. Find $P(0.5 \leq X \leq 2)$.

12. $f(x) = 0.5 - 0.125x$, $0 \leq x \leq 4$. Find $P(2 \leq X \leq 3)$.

13. $f(x) = 1/x$, $1 \leq x \leq e$. Find $P(1 \leq X \leq 2)$.

◣ 14. $f(x) = 6x(1 - x)$, $0 \leq x \leq 1$. Find $P(0.2 \leq X \leq 0.5)$.

15. Show that the function $F(x) = 1 - e^{-2x}$, $x \in [0, \infty)$, satisfies all properties listed in Theorem 13. Thus, it is a cumulative distribution function of a continuous random variable X. Find the corresponding probability density function and the expected value of X.

16. Show that $F(x) = 1 - x^{-3}$, $x \in [1, \infty)$, satisfies all properties listed in Theorem 13. Thus, it is a cumulative distribution function of a random variable X. Find the corresponding probability density function and the variance of X.

▽ 17–22 ▪ In each case:

(a) Check that $f(x)$ satisfies properties (1) and (2) in Definition 36.

(b) Find the cumulative distribution function for the distribution $f(x)$.

(c) Let X be a continuous random variable whose probability density function is $f(x)$. Find the expected value μ of X.

(d) Find the probability $P(X \le \mu)$.

17. $f(x) = 2x$, $0 \le x \le 1$

18. $f(x) = 8x$, $0 \le x \le 1/2$

19. $f(x) = 3x^2$, $0 \le x \le 1$

20. $f(x) = 4x^3$, $0 \le x \le 1$

▲ 21. $f(x) = \dfrac{2}{3} - \dfrac{2x}{9}$, $0 \le x \le 3$

22. $\dfrac{3}{4}x(2-x)$, $0 \le x \le 2$

23. Consider the continuous random variable X given by the probability density function $f(x) = 3x^2$, $0 \le x \le 1$. Find the probability that the values of X are at most one standard deviation away from the mean.

24. Consider the continuous random variable X given by the probability density function $f(x) = 0.3 + 0.2x$, $0 \le x \le 2$. Find $P(0.5 \le X \le 2)$. Find the probability that the values of X are at least one standard deviation above the mean.

▽ 25–28 ▪ Given the probability density function of a random variable X, answer each question.

25. $f(x) = 3x^2$, $0 \le x \le 1$. Find the median of X.

26. $f(x) = 3x^2$, $0 \le x \le 1$. Find the 95th percentile of X; i.e., find the value Q such that $P(X \le Q) = 0.95$.

27. $f(x) = \dfrac{2}{3} - \dfrac{2x}{9}$, $0 \le x \le 3$. Find the upper quartile of X; i.e., find Q_3 such that $P(X \le Q_3) = 0.75$.

▲ 28. $f(x) = 4x^3$, $0 \le x \le 1$. Find the lower quartile of X; i.e., find Q_1 such that $P(X \le Q_1) = 0.25$.

29. Suppose that the lifetime of a tree is given by the probability density function $f(t) = 0.01e^{-0.01t}$, where t is measured in years, $t \in [0, \infty)$. Find the average lifetime of the tree. What is the probability that it will live longer than 70 years?

30. Suppose that the lifetime of an insect is given by the probability density function $f(t) = 0.2e^{-0.2t}$, where t is measured in days, $t \in [0, \infty)$. What is the probability that the insect will live longer than 10 days?

31. The distance between a seed and the plant it came from is modelled by the density function

$$f(x) = \frac{2}{\pi(1 + x^2)}$$

where x represents the distance (in metres), $x \in [0, \infty)$. What is the probability that a seed will be found within 10 m of the plant?

32. The distance between a seed and the plant it came from is modelled by the density function

$$f(x) = \frac{2}{\pi(1 + x^2)}$$

where x represents the distance (in metres), $x \in [0, \infty)$. What is the probability that a seed will be found farther than 5 m from the plant?

33. Assume that $f(x) = 1 - |x|$, $-1 \leq x \leq 1$, is the probability density function of a random variable X.

    (a) Find $P(1/2 \leq X \leq 3/4)$ and $P(-1/2 \leq X \leq 0)$.

    (b) Find the expected value and the variance of X.

34. We calculate the integrals that are used in Example 13.13.

    (a) Show that the improper integral

    $$\int_0^\infty ae^{-ax}\, dx = 1$$

    (b) Imitating the integration by parts done in Example 13.10, show that

    $$\int axe^{-ax}\, dx = -\frac{1}{a}e^{-ax} - xe^{-ax}$$

    (c) Calculating the limits involved as in Example 13.10, show that

    $$\left(-\frac{1}{a}e^{-ax} - xe^{-ax} \right)\Big|_0^\infty = 0 - \left(\frac{1}{a} \right) = \frac{1}{a}$$

35. The Intermediate Value Theorem guarantees that a function takes on certain values. State the assumptions of the theorem, and show that they are fulfilled in the case of a cumulative distribution function $F(x)$ defined on an interval $[a, b]$, where a and b are real numbers. The function $F(x)$ defines the continuous random variable X. Conclude that the median of X exists.

14	The Normal Distribution

The **normal distribution** is the most important continuous distribution. It can be used to describe a variety of phenomena in biology, such as plant height, human growth or body weight distributions, the concentration of a chemical diffusing within a cell, litter size, the length of human or animal pregnancies, the size of animals, and so on. We will explain in what sense the normal distribution describes the outcomes of a large number of repetitions of a single experiment.

The graph of the probability density function of the normal distribution is a bell-shaped curve, also known as the *Gaussian distribution*.

The Normal Distribution

We start with a few examples and then define the probability density function for the normal distribution.

Example 14.1   Bacterial Population Dynamics Interpreted as Repeated Experiments

A bacterial culture contains 1,000 bacteria. Every day, each bacterium produces two offspring. The chance that both survive is 35%, the chance that one of them survives is 25%, and the chance that neither survives is 40%.

Denote by B_i ($i = 1, 2, \ldots, 1,000$) the random variable that counts the number of surviving offspring from bacterium i. The distribution of B_i is in Table 14.1.

Table 14.1

Surviving offspring	Probability
$P(B_i = 0)$	0.4
$P(B_i = 1)$	0.25
$P(B_i = 2)$	0.35

Since all B_i have the same distribution, they are called *identically distributed* random variables. The expected value of each B_i is

$$E(B_i) = (0)(0.4) + (1)(0.25) + (2)(0.35) = 0.95$$

and the variance is

$$\text{var}(B_i) = E(B_i^2) - (E(B_i))^2$$
$$= [(0)(0.4) + (1)(0.25) + (4)(0.35)] - 0.95^2 = 0.7475$$

Assume that the B_i are mutually independent and denote by B the total number of offspring in the next generation (one day later):

$$B = \sum_{i=1}^{1,000} B_i$$

We calculate the expected number of offspring using Theorem 7 in Section 7:

$$E(B) = \sum_{i=1}^{1,000} E(B_i) = 1,000(0.95) = 950$$

The variance is (here we need independence; see Theorem 9 in Section 9)

$$\text{var}(B) = \sum_{i=1}^{1,000} \text{var}(B_i) = 1,000(0.7475) = 747.5$$

So we know something about the distribution of B. But there are questions we might want to know the answers to, which cannot be obtained from the information we have here (or are not easy to get). What is the probability that the population will increase; i.e., what is the probability that $B > 1,000$? How likely is it that the population in the next generation will be between 900 and 1,000?

Once we learn how the normal distribution is related to the sum of independent, identically distributed random variables, we will be able to answer these questions.   ◣

Example 14.2   Questions We Will Answer Using the Normal Distribution

Numerous phenomena can be modelled using the normal distribution. By the end of this section we will be able to answer, among others, the following questions:

(a) How likely is it that an adult female Canadian is taller than 175 cm?

(b) The mean length of a human pregnancy from conception to birth is 266 days (this seems to be a widely accepted ballpark figure in North America; checking various sources, we found values of 272, 268, and 274 days; the common estimate of 40 weeks = 280 days is the duration measured from the date of the mother's last menstrual period; see, for instance Mittendorf, R., Williams, M.A., Berkey, C.S., & Cotter, P.F. (1990). The length of uncomplicated human gestation. *Obstetrics & Gynecology*, 75 (6), 929-932. or Omigbodun, A.O. (1997). Duration of human singleton pregnancies in Ibadan, Nigeria. *Journal of the National Medical Association*, 89 (9), 617-621).

A baby is considered premature if it is born less than 35 weeks from the date of conception. What is the probability of a premature birth? What is the length of 5% of the longest pregnancies? Do we have enough information to answer these questions? If not, what else do we need?

(c) The Wechsler Adult Intelligence Scale (WAIS) intelligence quotient (IQ) is distributed normally (bell-shaped) with a mean of 100 and a standard deviation of 15 (soon, we will learn what "normally distributed" means).

What is the IQ of the smartest 10% of the people, judging solely by the WAIS IQ test scores?   ◣

Example 14.3   Occurrence of a Virus

In Section 10 we studied the occurrence of a virus whose behaviour during month i is identified by the Bernoulli experiment

$$V_i = \begin{cases} 1 & \text{virus is present (success)} \\ 0 & \text{virus is absent} \end{cases}$$

where $P(V_i = 1) = 0.2$ and $P(V_i = 0) = 0.8$ for $i = 1, 2, \ldots, 120$. The random variable

$$N = \sum_{i=1}^{120} V_i$$

counts the number of months in a 10-year period during which the virus is present in a population. Its histogram (Figure 14.1) suggests a bell-shaped distribution.

What is the probability that the virus will be present for between 30 and 36 months during the 10-year period?

Of course, we can use the binomial distribution to answer this question, but the calculations are quite involved (see Exercise 39). In this section, we will learn how to approximate the binomial distribution with the normal distribution. This fact will make answering the question somewhat easier.   ◣

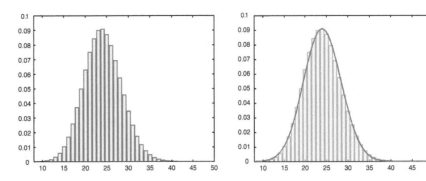

FIGURE 14.1

Histogram of the occurrence of the virus and its approximation

Definition 39   The Normal Distribution

We say that a continuous random variable X has a *normal distribution* (or is *distributed normally*) with mean μ and variance σ^2, and write $X \sim N(\mu, \sigma^2)$, if its probability density function is

$$f(x) = \frac{1}{\sigma\sqrt{2\pi}}\, e^{-(x-\mu)^2/2\sigma^2} \tag{14.1}$$

where x is in $(-\infty, \infty)$.

The probability density function of the normal distribution is determined by two parameters, $\mu \in \mathbb{R}$ and $\sigma > 0$. The range of X is $(-\infty, \infty)$.

In a moment we will describe the graph of $f(x)$. First, we justify the use of the terms "probability density function," "mean," and "variance" in Definition 39. Keep in mind that the second parameter in $X \sim N(\mu, \sigma^2)$ is the variance. The standard deviation $\sigma > 0$ is its square root.

Looking at (14.1), we see that $f(x) \geq 0$ for all x (since $\sigma > 0$). In order to verify that $f(x)$ is indeed a probability density function, we have to show that

$$\int_{-\infty}^{\infty} f(x)\, dx = \int_{-\infty}^{\infty} \frac{1}{\sigma\sqrt{2\pi}}\, e^{-(x-\mu)^2/2\sigma^2}\, dx = 1 \tag{14.2}$$

Using techniques that are beyond what we can do here one can prove that

$$\int_{-\infty}^{\infty} e^{-x^2}\, dx = \sqrt{\pi} \tag{14.3}$$

We can show that formula (14.2) follows from (14.3) using integration by substitution (see Exercise 40).

The mean of X is

$$E(X) = \int_{-\infty}^{\infty} x f(x)\, dx = \int_{-\infty}^{\infty} x \frac{1}{\sigma\sqrt{2\pi}}\, e^{-(x-\mu)^2/2\sigma^2}\, dx$$

Rewrite the exponent of e as $-\left(\frac{x-\mu}{\sigma\sqrt{2}}\right)^2$ and let $u = \frac{x-\mu}{\sigma\sqrt{2}}$. Then $x = \sqrt{2}\sigma u + \mu$, $du/dx = 1/\sqrt{2}\sigma$, and

$$E(X) = \frac{1}{\sigma\sqrt{2\pi}} \int_{-\infty}^{\infty} x e^{-(x-\mu)^2/2\sigma^2}\, dx$$

$$= \frac{1}{\sigma\sqrt{2\pi}} \int_{-\infty}^{\infty} \left(\sqrt{2}\sigma u + \mu\right) e^{-u^2} \sqrt{2}\sigma\, du$$

$$= \frac{1}{\sqrt{\pi}} \left(\int_{-\infty}^{\infty} \sqrt{2}\sigma u\, e^{-u^2}\, du + \int_{-\infty}^{\infty} \mu e^{-u^2}\, du \right)$$

$$= \frac{1}{\sqrt{\pi}} \left(\sqrt{2}\sigma \int_{-\infty}^{\infty} u\, e^{-u^2}\, du + \mu \int_{-\infty}^{\infty} e^{-u^2}\, du \right)$$

The first integral is zero (see Exercise 41) and the second is $\sqrt{\pi}$, by (14.3). Thus,

$$E(X) = \frac{1}{\sqrt{\pi}} \left(0 + \mu\sqrt{\pi} \right) = \mu$$

i.e., μ is indeed the expected value of X. The variance of X is given by

$$\mathrm{var}(X) = \frac{1}{\sigma\sqrt{2\pi}} \int_{-\infty}^{\infty} (x-\mu)^2 e^{-(x-\mu)^2/2\sigma^2} \, dx$$

In Exercise 42 we show that $\mathrm{var}(X) = \sigma^2$.

Now we describe the graph of the probability density function of the normal distribution.

Theorem 14   Properties of the Normal Distribution Density Function

The probability density function of the normal distribution

$$f(x) = \frac{1}{\sigma\sqrt{2\pi}} e^{-(x-\mu)^2/2\sigma^2}$$

satisfies the following properties:

(a) $f(x)$ is symmetric with respect to the vertical line $x = \mu$.

(b) $f(x)$ is increasing for $x < \mu$ and decreasing for $x > \mu$. It has a local (also global) maximum value $1/(\sigma\sqrt{2\pi})$ at $x = \mu$.

(c) The inflection points of $f(x)$ are $x = \mu - \sigma$ and $x = \mu + \sigma$.

(d) $\lim\limits_{x \to -\infty} f(x) = \lim\limits_{x \to \infty} f(x) = 0$

See Exercises 43 and 44 for proofs of these facts.

Figure 14.2 shows the graph of the probability density function for $N(10, 4)$, i.e., for the normal distribution with mean $\mu = 10$ and standard deviation $\sigma = 2$.

Note that the inflection points are one standard deviation below and above the mean, at $x = 10 - 2 = 8$ and $x = 10 + 2 = 12$. The maximum value the density function reaches is $1/(2\sqrt{2\pi}) \approx 0.1995$.

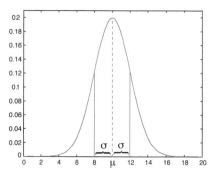

FIGURE 14.2

Normal distribution $N(10, 4)$

Assume that a continuous random variable X is normally distributed with mean μ and standard deviation σ; i.e., $X \sim N(\mu, \sigma^2)$. Then we calculate the probability $P(a \le X \le b)$ using

$$P(a \le X \le b) = \int_a^b f(x) \, dx = \int_a^b \frac{1}{\sigma\sqrt{2\pi}} e^{-(x-\mu)^2/2\sigma^2} \, dx \qquad (14.4)$$

This integral cannot be evaluated using elementary functions. It can be evaluated, for instance, by approximating the exponential function with a Taylor polynomial (see Appendix to this section). The usual approach to calculating the integral in (14.4) consists of reducing a general normal distribution to a special normal distribution (which we now define) and then using tables.

Definition 40   Standard Normal Distribution

The *standard normal distribution* is the normal distribution with mean 0 and variance 1; in symbols, it is $N(0, 1)$. Its probability density function is given by

$$f(x) = \frac{1}{\sqrt{2\pi}}\, e^{-x^2/2} \tag{14.5}$$

for all $x \in \mathbb{R}$.

Of course, (14.5) is obtained by substituting $\mu = 0$ and $\sigma = 1$ into (14.1). The graph of $f(x)$ is drawn in Figure 14.3.

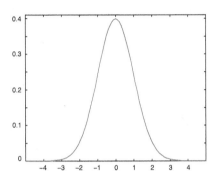

FIGURE 14.3

Density function for the standard normal distribution

We use the symbol Z to denote the continuous random variable that has the standard normal distribution; i.e., $Z \sim N(0, 1)$. The cumulative distribution function of Z is given by

$$F(z) = \int_{-\infty}^{z} f(x)\, dx = \int_{-\infty}^{z} \frac{1}{\sqrt{2\pi}}\, e^{-x^2/2}\, dx \tag{14.6}$$

$F(z)$ is a non-zero number, equal to the area of the shaded region in Figure 14.4a.

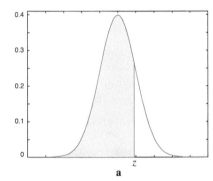
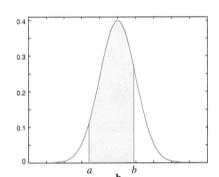

FIGURE 14.4

The area representing the cumulative distribution function and the probability

We give several values of $F(z)$ in Table 14.2. A larger set of values can be found in Table 14.4 in the Appendix at the end of this section.

Table 14.2

z	$F(z)$	z	$F(z)$
-4	0.000032	1	0.841345
-3	0.001350	2	0.977250
-2	0.022750	3	0.998650
-1	0.158655	4	0.999968
0	0.500000	5	0.999999

If $Z \sim N(0, 1)$, then

$$P(a \leq Z \leq b) = \int_a^b \frac{1}{\sqrt{2\pi}} e^{-x^2/2} \, dx = F(b) - F(a) \qquad (14.7)$$

This probability is equal to the area of the shaded region in Figure 14.4b.

Example 14.4   **Calculating Probabilities Using the Standard Normal Distribution**

Let $Z \sim N(0, 1)$. The probability that Z is less than 1 is (see Table 14.2)

$$P(Z < 1) = P(Z \leq 1) = \int_{-\infty}^1 \frac{1}{\sqrt{2\pi}} e^{-x^2/2} \, dx = F(1) = 0.841345$$

This probability is equal to the area of the shaded region in Figure 14.5a. The probability that Z is less than -3 is

$$P(Z < -3) = F(-3) = 0.001350$$

To calculate $P(Z \geq a)$, we use complementary events. For instance,

$$P(Z \geq 1) = 1 - P(Z < 1) = 1 - 0.841345 = 0.158655$$

(see Figure 14.5b). Likewise,

$$P(Z > 3) = 1 - P(Z \leq 3) = 1 - 0.998650 = 0.001350 \qquad (14.8)$$

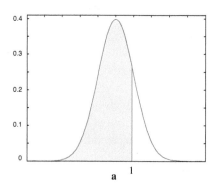 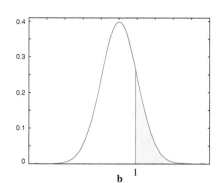

FIGURE 14.5

Areas representing probabilities

The result in (14.8) means that very few values (about 0.14%) of Z are more than three standard deviations above the mean (keep in mind that $Z \sim N(0, 1)$, so the mean is 0 and the standard deviation is 1). The fact that $P(Z \geq 1) = 0.158655$ means that a bit over 15.8% of the values of Z are more than one standard deviation above the mean.

Using (14.7), we find that

$$P(-2 \leq Z \leq 1) = \int_{-2}^1 \frac{1}{\sqrt{2\pi}} e^{-x^2/2} \, dx$$

$$= F(1) - F(-2) = 0.841345 - 0.022750 = 0.818595$$

Thus, the area of the shaded region in Figure 14.6 is 0.818595.

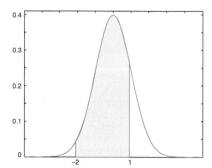

FIGURE 14.6

Probability as area

Next, we compute three probabilities over symmetric intervals.

$$P(-1 \leq Z \leq 1) = \int_{-1}^{1} \frac{1}{\sqrt{2\pi}} e^{-x^2/2} \, dx$$
$$= F(1) - F(-1)$$
$$= 0.841345 - 0.158655 = 0.682690$$

In words, a bit over 68% of the values of Z fall within one standard deviation of the mean. Likewise,

$$P(-2 \leq Z \leq 2) = F(2) - F(-2) = 0.977250 - 0.022750 = 0.954500$$

and

$$P(-3 \leq Z \leq 3) = F(3) - F(-3) = 0.998650 - 0.001350 = 0.997300$$

Thus, about 95.5% of the values of Z fall within two standard deviations of the mean, and about 99.7% are within three standard deviations of the mean.

The conclusion we reached in Example 14.4 holds in general: if $X \sim N(\mu, \sigma^2)$, i.e., if X is a continuous random variable distributed normally with mean μ and standard deviation σ, then

$$\begin{aligned} P(\mu - \sigma \leq X \leq \mu + \sigma) &= 0.683 \\ P(\mu - 2\sigma \leq X \leq \mu + 2\sigma) &= 0.955 \\ P(\mu - 3\sigma \leq X \leq \mu + 3\sigma) &= 0.997 \end{aligned} \qquad (14.9)$$

(see Exercise 45). The formulas in (14.9) are referred to as the "68-95-99.7 rule." In words, for a normally distributed random variable:

68.3% of the values fall within one standard deviation of the mean.

95.5% of the values fall within two standard deviations of the mean.

99.7% of the values fall within three standard deviations of the mean.

See Figure 14.7.

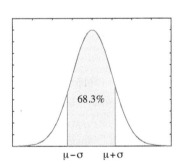
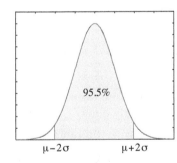
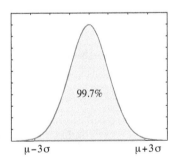

FIGURE 14.7

The distribution of the values of a normal distribution

Example 14.5   Finding Probabilities Using the 68-95-99.7 Rule

Assume that $X \sim N(\mu, \sigma^2)$. Using (14.9), find

(a) $P(X < \mu + \sigma)$

(b) $P(X \leq \mu + 3\sigma)$

(c) $P(X > \mu + \sigma)$

▶ (a) We are asked to find the area of the shaded region in Figure 14.8a. Since the probability density function is symmetric with respect to the mean μ, the area of the region to the left of μ is 0.5. We know that 68.3% of the values of X lie between $\mu - \sigma$ and $\mu + \sigma$. Again, due to the symmetry of the graph, one half of these values, $68.3\%/2 = 34.15\%$, lie between μ and $\mu + \sigma$. Thus,

$$P(X < \mu + \sigma) = P(X \leq \mu + \sigma) = 0.5 + 0.3415 = 0.8415$$

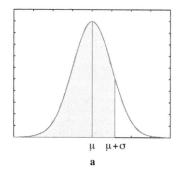
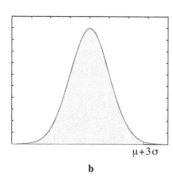
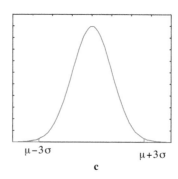

a   b   c

FIGURE 14.8

Calculating probability using areas

(b) We know that 99.7% of the values of X lie between $\mu - 3\sigma$ and $\mu + 3\sigma$. Thus, $99.7\%/2 = 49.85\%$ of the values lie between μ and $\mu + 3\sigma$. It follows that

$$P(X \leq \mu + 3\sigma) = P(X \leq \mu) + P(\mu \leq X \leq \mu + 3\sigma) = 0.5 + 0.4985 = 0.9985$$

See Figure 14.8b.

Alternatively, we argue in the following way: the probability that X is farther than three standard deviations from the mean is $1 - 0.997 = 0.003$. That's the area of the two "tails" of the distribution in Figure 14.8c. By symmetry, the area of each tail is $0.003/1 = 0.0015$. Thus,

$$P(X \leq \mu + 3\sigma) = 1 - \text{area of the tail at the right end}$$
$$= 1 - 0.0015 = 0.9985$$

(c) Using (a), we get

$$P(X > \mu + \sigma) = 1 - P(X \leq \mu + \sigma) = 1 - 0.8415 = 0.1585.$$

Example 14.6 **Using the 68-95-99.7 Rule**

Assume that a certain quantity X is distributed normally, $X \sim N(15, 3^2)$.

(a) Find an interval centred at the mean of X with the property that there is about a 95% chance that a randomly chosen value of X falls into this interval.

(b) Estimate the probability that a randomly chosen value of X is larger than 24.

▶ The variance of X is $3^2 = 9$, so its standard deviation is $\sigma = 3$.

(a) Looking at (14.9), we see that the range of values corresponding to a 95% chance is within two standard deviations of the mean, so the interval is $[15 - 2 \cdot 3, 15 + 2 \cdot 3] = [9, 21]$.

(b) Note that $24 = 15 + 3 \cdot 3$, i.e., 24 is three standard deviations of the mean. The probability that $X > 24$ is equal to the area of the right tail in Figure 14.8c, which is 0.0015.

Example 14.7 **The Lengths of Pregnancies**

The lengths of human pregnancies (measured in days from conception to birth) can be approximated by the normal distribution with a mean of 266 days and a standard deviation of 16 days (see Example 14.2(b) for references).

Thus, about 68% of pregnancies last between $266 - 16 = 250$ days and $266 + 16 = 282$ days. About 95.5% of pregnancies last between $266 - 2 \cdot 16 = 234$ days and $266 + 2 \cdot 16 = 298$ days, and about 99.7% of pregnancies last between $266 - 3 \cdot 16 = 218$ days and $266 + 3 \cdot 16 = 314$ days.

Assume that $X \sim N(\mu, \sigma^2)$. If $a, b \in \mathbb{R}$, then

$$P(a \leq X \leq b) = \int_a^b \frac{1}{\sigma\sqrt{2\pi}} e^{-(x-\mu)^2/2\sigma^2} \, dx$$

Using the substitution $z = (x - \mu)/\sigma$ and $dz/dx = 1/\sigma$, we obtain

$$\int_a^b \frac{1}{\sigma\sqrt{2\pi}} e^{-(x-\mu)^2/2\sigma^2} \, dx = \int_{(a-\mu)/\sigma}^{(b-\mu)/\sigma} \frac{1}{\sigma\sqrt{2\pi}} e^{-z^2/2} \, \sigma \, dz$$

$$= \int_{(a-\mu)/\sigma}^{(b-\mu)/\sigma} \frac{1}{\sqrt{2\pi}} e^{-z^2/2} \, dz$$

The right side is the probability

$$P\left(\frac{a-\mu}{\sigma} \leq Z \leq \frac{b-\mu}{\sigma}\right)$$

and therefore

$$P(a \leq X \leq b) = P\left(\frac{a-\mu}{\sigma} \leq Z \leq \frac{b-\mu}{\sigma}\right) \tag{14.10}$$

where Z is the standard normal distribution.

In other words, the area under the normal distribution density function between a and b is equal to the area under the standard normal distribution density function between $(a - \mu)/\sigma$ and $(b - \mu)/\sigma$. Thus, the two shaded regions in Figure 14.9 have the same area.

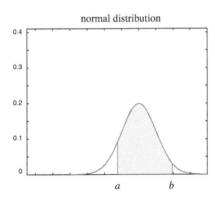
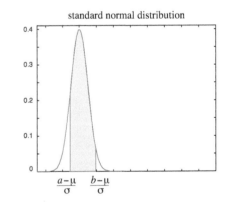

FIGURE 14.9

The normal and the standard normal distributions

Thus, we have proved the following fact.

Theorem 15 **The Normal and the Standard Normal Distributions**

Assume that $X \sim N(\mu, \sigma^2)$. The random variable $Z = (X - \mu)/\sigma$ has the standard normal distribution, i.e., $Z \sim N(0, 1)$.

Putting it all together, if $X \sim N(\mu, \sigma^2)$, then

$$P(a \leq X \leq b) = P\left(\frac{a-\mu}{\sigma} \leq Z \leq \frac{b-\mu}{\sigma}\right)$$

$$= F\left(\frac{b-\mu}{\sigma}\right) - F\left(\frac{a-\mu}{\sigma}\right) \tag{14.11}$$

where F is the cumulative distribution function of the standard normal distribution $Z \sim N(0,1)$. We read the values of F from a table (such as Table 14.4 in the Appendix to this section).

One technical issue: Table 14.4 lists the values $F(z)$ for positive z only. How do we calculate $F(-0.75)$ or $F(-4)$?

Example 14.8   **Calculating the Values of $F(z)$ for Negative z**

It is given that $F(1.5) = 0.933193$. Find $F(-1.5)$.

▶ Let's use pictures. Denote by A, B, and C the areas of the three shaded regions in Figure 14.10. We are asked to find A (or C, since they are equal by the symmetry of the graph).

It is given that $A + B = F(1.5)$. Since $A + B + C = 1$, it follows that

$$F(-1.5) = A = C$$
$$= 1 - (A + B)$$
$$= 1 - F(1.5)$$
$$= 1 - 0.933193 = 0.066807$$

So we have discovered the formula $F(-1.5) = 1 - F(1.5)$.

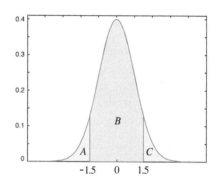

FIGURE 14.10

Computing $F(-1.5)$

Replacing 1.5 by z and -1.5 by $-z$ in Example 14.8, we obtain

$$F(-z) = 1 - F(z) \tag{14.12}$$

for any z. As practice, we now verify (14.12) using probability density functions. We know $F(z)$, $z \geq 0$, and need to find

$$F(-z) = \int_{-\infty}^{-z} \frac{1}{\sqrt{2\pi}} e^{-x^2/2} \, dx$$

By the symmetry of the graph ($A = C$ in Figure 14.10) we get

$$F(-z) = \int_{-\infty}^{-z} \frac{1}{\sqrt{2\pi}} e^{-x^2/2} \, dx = \int_{z}^{\infty} \frac{1}{\sqrt{2\pi}} e^{-x^2/2} \, dx$$

Since

$$1 = \int_{-\infty}^{\infty} \frac{1}{\sqrt{2\pi}} e^{-x^2/2} \, dx = \int_{-\infty}^{z} \frac{1}{\sqrt{2\pi}} e^{-x^2/2} \, dx + \int_{z}^{\infty} \frac{1}{\sqrt{2\pi}} e^{-x^2/2} \, dx$$

we obtain

$$\int_{z}^{\infty} \frac{1}{\sqrt{2\pi}} e^{-x^2/2} \, dx = 1 - \int_{-\infty}^{z} \frac{1}{\sqrt{2\pi}} e^{-x^2/2} \, dx$$

i.e.,

$$F(-z) = 1 - F(z)$$

and we are done.

In particular,
$$F(-0.75) = 1 - F(0.75) = 1 - 0.773373 = 0.226627 \qquad (14.13)$$
and
$$F(-4) = 1 - F(4) = 1 - 0.999968 = 0.000032 \qquad (14.14)$$

Now we are ready to calculate the probability $P(a \leq X \leq b)$ for the normal distribution $X \sim N(\mu, \sigma)$.

According to (14.11), we convert a and b by subtracting the mean and dividing by the standard deviation. The values $(a - \mu)/\sigma$ and $(b - \mu)/\sigma$ are called the *z-scores* of a and b, respectively. We think of the z-score of a number as the difference from that number to the mean, in units of standard deviations.

Example 14.9 Calculating Probabilities

Assume that $X \sim N(25, 16)$. Find $P(22 \leq X \leq 30)$.

▶ The random variable X is normally distributed with mean $\mu = 25$ and standard deviation $\sigma = 4$. To convert $P(22 \leq X \leq 30)$ to the probability related to the standard normal distribution, we calculate the z-scores; the z-score of 22 is
$$\frac{22 - \mu}{\sigma} = \frac{22 - 25}{4} = -\frac{3}{4} = -0.75$$
and the z-score of 30 is
$$\frac{30 - \mu}{\sigma} = \frac{30 - 25}{4} = \frac{5}{4} = 1.25$$
Thus,
$$P(22 \leq X \leq 30) = P(-0.75 \leq Z \leq 1.25)$$
where $Z \sim N(0, 1)$. Using Table 14.4 from the Appendix and (14.13),
$$P(-0.75 \leq Z \leq 1.25) = F(1.25) - F(-0.75)$$
$$= 0.894350 - 0.226627 = 0.667723$$

Likewise, using (14.14) and Table 14.4,
$$P(9 \leq X \leq 29) = P\left(\frac{9 - 25}{4} \leq Z \leq \frac{29 - 25}{4}\right)$$
$$= P(-4 \leq Z \leq 1)$$
$$= F(1) - F(-4)$$
$$= 0.841345 - 0.000032 = 0.841313$$

We are ready to answer the questions we asked at the beginning of this section.

Example 14.10 Answers to Questions from Example 14.2

(a) Human height data have been collected for a variety of reasons. For instance, there is a growing body of literature that explores the relationship between the mean human height in a population and the living standard. In some cases, the normal distribution is used to model human height, although a number of sources that use it admit that the symmetry of the normal distribution is too restrictive to reflect the true distribution.

For our example, we will, nevertheless, assume that the distribution of adult female (age group 25–44) heights in Canada is normal. Taking the mean to be 163 cm and the standard deviation to be 6 cm, we consider the distribution of heights $H \sim N(163, 6^2)$. [Source: Shields, M., Connor Gorber, S., & Tremblay, M.S. (2008). Methodological issues in anthropometry: Self-reported versus measured height and weight. *Proceedings of Statistics Canada Symposium 2008.* Available at www.statcan.gc.ca/pub/11-522-x/2008000/article/11002-eng.pdf.]

Thus, the probability that a female Canadian (in the age range 25–44) is taller than 172 cm is

$$P(H \geq 172) = P\left(\frac{H - 163}{6} \geq \frac{172 - 163}{6}\right)$$
$$= P(Z \geq 1.5)$$
$$= 1 - P(Z < 1.5)$$
$$= 1 - F(1.5) = 1 - 0.933193 = 0.066807$$

i.e., about 6.7%.

(b) It is given that the mean length of a human pregnancy from conception to birth is 266 days. In order to use the normal distribution, we need the standard deviation; from the sources listed in Example 14.2(b), we learn that $\sigma = 16$. Thus, we model the length of pregnancy as the random variable $L \sim N(266, 16^2)$.

The probability that a baby is born prematurely is (35 weeks is 245 days)

$$P(L \leq 245) = P\left(\frac{L - 266}{16} \leq \frac{245 - 266}{16}\right)$$
$$= P(Z \leq -1.3125)$$
$$= F(-1.3125)$$

We do not have the value 1.3125 in Table 14.4, so we use the nearest one: $F(1.3) = 0.903202$. Thus,

$$F(-1.3125) \approx F(-1.3) = 1 - F(1.3) = 1 - 0.903200 = 0.096800$$

In words, about 10% of babies are born prematurely.

What is the length of the longest 5% of pregnancies? To answer this question, we need to find ℓ such that $P(L > \ell) = 0.05$ or $P(L \leq \ell) = 0.95$. We start as usual:

$$P(L \leq \ell) = 0.95$$
$$P\left(\frac{L - 266}{16} \leq \frac{\ell - 266}{16}\right) = 0.95$$

We note that this is an "inverse" question: we know the probability, and need to figure out the z-score. Looking at Table 14.4, we read

$$P(Z \leq 1.65) = 0.950529 \approx 0.95$$

Comparing the two expressions, we get

$$\frac{\ell - 266}{16} = 1.65$$
$$\ell = 266 + (16)(1.65) = 292.40$$

days (which is 41 weeks, 5 days and about 10 hours).

(c) It is given that the WAIS IQ test scores distribution is $T \sim N(100, 15^2)$. We are asked to determine the IQ of the smartest 10% of the people judging solely by the WAIS IQ test scores.

As in (b), this is an inverse question: we need to find t so that $P(T > t) = 0.1$ or, equivalently, $P(T \leq t) = 0.9$:

$$P(T \leq t) = 0.9$$
$$P\left(\frac{T - 100}{15} \leq \frac{t - 100}{15}\right) = 0.9$$

In Table 14.4, we find $F(1.3) = 0.903200$ (close enough to 0.9). Thus

$$\frac{t - 100}{15} = 1.3$$
$$t = 100 + (15)(1.3) = 119.5$$

So, the smartest 10% of the people have a WAIS IQ of 119.5 or higher.   ◣

Looking back at Examples 14.2 and 14.10, how was it determined that the mean length of pregnancy is 266 days, or that the average height of an adult Canadian female is 163 cm?

It is impossible to collect measurements from the whole population. No matter what is investigated (statistically), researchers need to pick a *sample* that is *representative* of the whole population. How exactly this is done (or whether it can be done at all) is a difficult problem that we do not discuss here.

In any case, based on a sample and *assuming* that the phenomenon under investigation has a certain distribution, statisticians estimate the parameters for the distribution. The pregnancy length and the female height distributions are based on data collected from a sample. Under the assumption that the distributions are normal, the estimates for μ and σ are derived.

To understand the type of difficulty statisticians face in going from a sample to a distribution, consider the following example.

Example 14.11   Sample versus Population

The following samples (representing some quantity) have been taken from the same population, assumed to be normally distributed with a mean of 10 and a standard deviation of 3:

$$S_1 = \{8, 9, 5, 9, 6, 10, 13, 11, 8, 5, 14, 16, 10, 7, 14, 8, 9, 13, 6, 4\}$$
$$S_2 = \{11, 11, 10, 11, 13, 13, 8, 12, 11, 8, 6, 12, 9, 10, 8, 5, 11, 10, 14, 12\}$$

(To get these values, we used a random number generator programmed to pick values according to the distribution $X \sim N(10, 3^2)$.)

Since the mean is 10, roughly one half of the measurements should be above and one half below 10. In sample S_1, 12 values are below 10, 6 values are above 10, and two values are equal to 10. In sample S_2, 6 values are below 10, 11 values are above 10, and 3 values are equal to 10.

As well, 68% of the values should be within one standard deviation of the mean, i.e., in the interval $[7, 13]$. In sample S_1, 12 values (60% of all values) belong to the interval. In sample S_2, 17 values (85%) are in the interval $[7, 13]$.

Thus, in spite of the fact that they came from the same population, the two samples are far from identical.   ◣

Obviously, S_1 and S_2 differ. But if we increase their size (say, pick 50 or 200 values instead of 20), the samples will resemble each other more closely, and will represent the total population more faithfully. (We will not go into this issue any further.)

Central Limit Theorems

One reason we study the normal distribution is its ability to describe phenomena whose values depend on a large number of small contributions. For instance, small errors in rounding off a number add up as we perform large, complex calculations. The height and the weight of an animal are believed to be the result of various factors that are mutually independent and act additively (i.e., each contributes a small quantity to the height or to the weight).

Consider the situation in Example 14.1. Each of the 1,000 bacteria can be viewed as an experiment in which we measure the number of surviving offspring. The outcome of each experiment is small compared to the total population and contributes *additively* to the total population of bacteria in the next generation (a day later). It is assumed that the number of surviving offspring of one bacterium is independent of all other bacteria, and that all bacteria have the same probability distribution for the survival of their offspring.

Define B_i = "number of surviving offspring from bacterium i," where $i = 1, 2, \ldots, 1{,}000$. Keep in mind that all B_i are mutually independent and *identically distributed* (all have the same probability distribution). We are interested in their sum

$$B = B_1 + B_2 + \cdots + B_{1{,}000} = \sum_{i=1}^{1{,}000} B_i$$

since it gives the number of bacteria in the next generation.

Theorem 16   **Central Limit Theorem for Sums of Random Variables**

Assume that the random variables $X_1, X_2, \ldots, X_n$ are mutually independent and identically distributed, with mean μ and variance σ^2. Define the random variable

$$S = \sum_{i=1}^{n} X_i$$

For sufficiently large n, the probability density function of S can be approximated by the normal distribution with mean $n\mu$ and variance $n\sigma^2$.   ◢

The proof of this theorem is beyond the scope of this book.

Note that the random variables $X_1, X_2, \ldots, X_n$ do not have to be normally distributed; all we require is that they be identically distributed and independent.

What is the "sufficiently large" value of n in Theorem 16? It is generally accepted that the theorem will give meaningful results when $n \geq 30$.

In our example with the bacteria, $n = 1{,}000$. Let's go back to it.

Example 14.12   **Answering Questions from Example 14.1**

Consider the sum

$$B = \sum_{i=1}^{1{,}000} B_i$$

that gives the number of offspring in the next generation. We assume that the B_i are mutually independent. In Example 14.1 we calculated $E(B_i) = 0.95$ and $\mathrm{var}(B_i) = 0.7475$ for all i. By the Central Limit Theorem,

$$B \sim N(n\mu, n\sigma^2) = N(950, 747.50)$$

The standard deviation of B is $\sqrt{747.50} \approx 27.34$.

The probability that the population will increase in the next generation is

$$P(B > 1{,}000) = 1 - P(B \leq 1{,}000)$$
$$= 1 - P\left(\frac{B - 950}{27.34} \leq \frac{1{,}000 - 950}{27.34}\right)$$
$$= 1 - P(Z \leq 1.83)$$
$$\approx 1 - P(Z \leq 1.85)$$
$$= 1 - 0.967843 = 0.032157$$

So, there is little chance (about 3.2%) that the population will increase in size.

How likely is it that in the next generation there will be between 900 and 1,000 bacteria?

Note that the range $[900, 1{,}000]$ is about two standard deviations from the mean; the exact interval is

$$[950 - (2)(27.34), 950 + (2)(27.34)] = [895.32, 1{,}004.68]$$

Thus, we expect the probability to be around 0.95. We compute

$$P(900 \le B \le 1{,}000) = P\left(\frac{900 - 950}{27.34} \le Z \le \frac{1{,}000 - 950}{27.34}\right)$$

$$= P(-1.83 \le Z \le 1.83)$$

$$= F(1.83) - F(-1.83)$$

$$= F(1.83) - (1 - F(1.83))$$

$$= 2F(1.83) - 1 \approx 2(0.967843) - 1 = 0.935686$$

We approximated $F(1.83)$ by $F(1.85)$ so that we can use Table 14.4.

The Central Limit Theorem talks about sums of random variables. The reason the sum of random variables can be approximated by the normal distribution is that the normal distribution is *additive*. More precisely, the sum $X = X_1 + X_2$ of the two independent random variables $X_1 \sim N(\mu_1, \sigma_1^2)$ and $X_2 \sim N(\mu_2, \sigma_2^2)$ is the normal distribution

$$X = X_1 + X_2 \sim N(\mu_1 + \mu_2, \sigma_1^2 + \sigma_2^2)$$

We check that

$$E(X) = E(X_1) + E(X_2) = \mu_1 + \mu_2$$

Because X_1 and X_2 are independent,

$$\mathrm{var}(X) = \mathrm{var}(X_1) + \mathrm{var}(X_2) = \sigma_1^2 + \sigma_2^2$$

Note that we did not prove that X is normally distributed (that's the difficult part). We only proved that the means and the variances add up.

Example 14.13   **The Occurrence of a Virus, Revisited**

We answer the question asked in Example 14.3.

The random variable $M = V_1 + V_2 + \cdots + V_{120}$ counts the number of months during which the virus is present (out of $n = 120$ months). Recall that $P(V_i = 1) = p = 0.2$ for each $i = 1, 2, \ldots, 120$. In Example 10.15 in Section 10 we calculated the values

$$E(V_i) = p = 0.2$$

$$\mathrm{var}(V_i) = p(1 - p) = (0.2)(0.8) = 0.16$$

The random variables V_i are identically distributed (and assumed to be) independent. The mean of M is (see Theorem 7 in Section 7)

$$E(M) = np = (120)(0.2) = 24$$

and the variance is (see Theorem 9 in Section 9)

$$\mathrm{var}(M) = np(1 - p) = 120(0.2)(0.8) = 19.2$$

Using the Central Limit Theorem, we approximate M by the normal distribution $M \sim N(24, 19.2)$.

To find the probability that the virus will be present in between 30 and 36 months during the 10-year period, we compute (the standard deviation is $\sqrt{19.2} \approx 4.38$)

$$P(30 \le M \le 36) = P\left(\frac{30 - 24}{4.38} \le Z \le \frac{36 - 24}{4.38}\right)$$

$$\approx P(1.37 \le Z \le 2.74)$$

$$= F(2.74) - F(1.37)$$

$$\approx 0.997020 - 0.911492 = 0.085528$$

To get the values from Table 14.4 we used the approximations $F(2.75)$ for $F(2.74)$ and $F(1.35)$ for $F(1.37)$.

In Exercise 39 we discover the difficulties that we encounter when we try to calculate this probability using the binomial distribution. Although it seems much simpler to do it using the normal distribution approximation, keep in mind that we did not do the hard part of the calculation—instead, we read the values of the cumulative distribution function F from a table. ◢

The close connection between the average and the sum is given in the following consequence of the Central Limit Theorem.

Theorem 17   Central Limit Theorem for Averages

Assume that the random variables $X_1, X_2, \ldots, X_n$ are mutually independent and identically distributed with mean μ and variance σ^2. Define the random variable

$$A = \frac{1}{n} \sum_{i=1}^{n} X_i$$

For sufficiently large n, the probability density function of A can be approximated by the normal distribution with mean μ and variance σ^2/n. ◢

As before, "sufficiently large n" means $n \geq 30$. Note that $A = \frac{1}{n}S$, where S is the sum random variable from Theorem 16. The statement of Theorem 17 is easy to verify:

$$E(A) = E\left(\frac{1}{n}S\right) = \frac{1}{n}E(S) = \frac{1}{n}n\mu = \mu$$

As well (because $X_1, X_2, \ldots, X_n$ are mutually independent),

$$\text{var}(A) = \text{var}\left(\frac{1}{n}S\right) = \frac{1}{n^2}\text{var}(S) = \frac{1}{n^2}n\sigma^2 = \frac{\sigma^2}{n}$$

Example 14.14   Average Number of Offspring of Bacteria

The number (see Example 14.12)

$$A = \frac{1}{1,000} \sum_{i=1}^{1,000} B_i$$

gives the average number of offspring in the next generation. According to Theorem 17, A is approximatelly normally distributed:

$$A \sim N(\mu, \sigma^2/n) = N(0.95, 0.7475/1,000) = N(0.95, 0.0007475) \quad ◢$$

Example 14.15   Population with Immigration and Emigration

We study the fluctuations in the size of a population of lions in a national park due to movements of lions into the region or out of it. The distribution of the net yearly change (= number of lions that immigrated minus number of lions that emigrated) is given in Table 14.3.

Assume that the same pattern persists for 40 years, and that these patterns are mutually independent. Find the normal distribution approximation for the total change in the population of lions due to immigration and emigration.

Table 14.3

Net yearly change in number of lions	Probability
−3	0.3
−1	0.15
0	0.2
2	0.15
6	0.2

▶ Denote by C_i the net yearly change in the lion population due to immigration and emigration, $i = 1, 2, \ldots, 40$. We are interested in the total change over 40 years:

$$C = \sum_{i=1}^{40} C_i$$

We compute

$$E(C_i) = (-3)(0.3) + (-1)(0.15) + (0)(0.2) + (2)(0.15) + (6)(0.2) = 0.45$$

for all i. From

$$E(C_i^2) = (-3)^2(0.3) + (-1)^2(0.15) + (0)^2(0.2) + (2)^2(0.15) + (6)^2(0.2) = 10.65$$

we compute the variance

$$\mathrm{var}(C_i) = E(C_i^2) - (E(C_i))^2 = 10.65 - 0.45^2 = 10.4475$$

for all i. Using the Central Limit Theorem (Theorem 16), we approximate C by the normal distribution

$$C \sim N(40 \cdot 0.45, 40 \cdot 10.4475) = N(18, 417.9)$$

The average total change $A = \frac{1}{40} \sum_{i=1}^{40} C_i$ is distributed as (Theorem 17)

$$A \sim N(0.45, 10.4475/40) \approx N(0.45, 0.26)$$

Appendix: Values of the Cumulative Distribution Function $F(z)$

In Table 14.4 we give the values of the cumulative distribution function

$$F(z) = \int_{-\infty}^{z} \frac{1}{\sqrt{2\pi}} e^{-x^2/2} \, dx$$

of the standard normal distribution for z between 0 and 4. The value $F(z)$ is equal to the area of the shaded region in Figure 14.11.

FIGURE 14.11

The area of the shaded region is $F(z)$

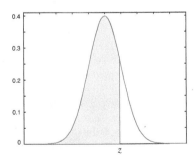

Remember to use

$$F(-z) = 1 - F(z)$$

to calculate the values of F for negative z.

Table 14.4

z	$F(z)$	z	$F(z)$	z	$F(z)$	z	$F(z)$
0	0.500000	1	0.841345	2	0.977250	3	0.998650
0.05	0.519938	1.05	0.853141	2.05	0.979818	3.05	0.998856
0.1	0.539828	1.1	0.864334	2.1	0.982136	3.1	0.999032
0.15	0.559618	1.15	0.874928	2.15	0.984222	3.15	0.999184
0.2	0.579260	1.2	0.884930	2.2	0.986097	3.2	0.999313
0.25	0.598706	1.25	0.894350	2.25	0.987776	3.25	0.999423
0.3	0.617911	1.3	0.903200	2.3	0.989276	3.3	0.999517
0.35	0.636831	1.35	0.911492	2.35	0.990613	3.35	0.999596
0.4	0.655422	1.4	0.919243	2.4	0.991802	3.4	0.999663
0.45	0.673645	1.45	0.926471	2.45	0.992857	3.45	0.999720
0.5	0.691462	1.5	0.933193	2.5	0.993790	3.5	0.999767
0.55	0.708840	1.55	0.939429	2.55	0.994614	3.55	0.999807
0.6	0.725747	1.6	0.945201	2.6	0.995339	3.6	0.999840
0.65	0.742154	1.65	0.950529	2.65	0.995975	3.65	0.999869
0.7	0.758036	1.7	0.955435	2.7	0.996533	3.7	0.999892
0.75	0.773373	1.75	0.959941	2.75	0.997020	3.75	0.999912
0.8	0.788145	1.8	0.964070	2.8	0.997445	3.8	0.999928
0.85	0.802337	1.85	0.967843	2.85	0.997814	3.85	0.999941
0.9	0.815940	1.9	0.971283	2.9	0.998134	3.9	0.999952
0.95	0.828944	1.95	0.974412	2.95	0.998411	3.95	0.999961
						4	0.999968

How were the values in Table 14.4 calculated?

One way to do it is to approximate $F(z)$ using Taylor polynomials. As an example, we show how to find $F(1)$. Note that

$$F(1) = \int_{-\infty}^{1} \frac{1}{\sqrt{2\pi}}\, e^{-x^2/2}\, dx$$

$$= \int_{-\infty}^{0} \frac{1}{\sqrt{2\pi}}\, e^{-x^2/2}\, dx + \int_{0}^{1} \frac{1}{\sqrt{2\pi}}\, e^{-x^2/2}\, dx$$

$$= 0.5 + \frac{1}{\sqrt{2\pi}} \int_{0}^{1} e^{-x^2/2}\, dx \tag{14.15}$$

Recall that we can approximate e^x using the Taylor polynomial

$$e^x \approx 1 + x + \frac{x^2}{2!} + \frac{x^3}{3!} + \cdots + \frac{x^n}{n!}$$

Replacing x by $-x^2/2$, we obtain

$$e^{-x^2/2} \approx 1 - \frac{x^2}{2} + \frac{(-x^2/2)^2}{2!} + \frac{(-x^2/2)^3}{3!} + \cdots + \frac{(-x^2/2)^n}{n!}$$

$$= 1 - \frac{x^2}{2} + \frac{x^4}{2^2 \cdot 2!} - \frac{x^6}{2^3 \cdot 3!} + \cdots + \frac{(-1)^n x^{2n}}{2^n n!} \tag{14.16}$$

To approximate the integral in (14.15), we use the first four terms (of course, increasing the number of terms improves the approximation):

$$\int_0^1 e^{-x^2/2}\,dx \approx \int_0^1 \left(1 - \frac{x^2}{2} + \frac{x^4}{8} - \frac{x^6}{48}\right) dx$$

$$= \left(x - \frac{x^3}{6} + \frac{x^5}{40} - \frac{x^7}{336}\right)\Bigg|_0^1$$

$$= 1 - \frac{1}{6} + \frac{1}{40} - \frac{1}{336} = 0.855357$$

Thus,

$$F(1) = 0.5 + \frac{1}{\sqrt{2\pi}} \int_0^1 e^{-x^2/2}\,dx \approx 0.5 + \frac{1}{\sqrt{2\pi}} 0.855357 = 0.841238$$

which agrees with $F(1) = 0.841347$ to three decimal places.

Summary The **normal distribution** can be used to model a wide variety of phenomena in the life sciences and elsewhere. The probability density function of the normal distribution is a bell-shaped curve, symmetric about the mean. In order to calculate the probability related to a normal distribution, we use the **standard normal distribution.** We convert the values to *z*-scores by subtracting the mean and dividing by the standard deviation. The calculation of the values of the cumulative distribution function of the standard normal distribution involves an integral that cannot be evaluated using elementary means. Instead, we use tables or mathematical software to find its values. The sum of independent, identically distributed random variables can be approximated by a normal distribution. This idea is made precise in the statement of the **Central Limit Theorem.**

14	Exercises

1. What is the *z*-score? Explain how to calculate the probability $P(0 \le X \le 7)$ if $X \sim N(3, 16)$.

2. Sketch the graph of the *standard* normal distribution. Shade the region whose area corresponds to the probability $P(1 \le X \le 4)$, if $X \sim N(3, 1^2)$.

3. Sketch the graph of the *standard* normal distribution. Shade the region whose area corresponds to the probability $P(-1 \le X \le 2)$, if $X \sim N(0, 2^2)$.

▽ 4–11 ▪ Find each probability using Table 14.4.

4. X is normally distributed with mean 3 and variance 4. Find the probability that X is less than 4.1.

5. X is normally distributed with mean 5 and variance 100. Find the probability that X is less than 9.

6. Let $X \sim N(-1, 4)$; find $P(X > 1)$.

7. Let $X \sim N(0, 10^2)$; find $P(X > 25)$.

8. Let $X \sim N(-1, 4)$; find $P(X < -2)$.

9. Let $X \sim N(-5, 10^2)$; find $P(X < -10)$.

10. Let $X \sim N(-2, 4)$; find $P(-3 \le X \le 1)$.

11. Let $X \sim N(2, 5^2)$; find $P(0 \le X \le 5)$.

12. The wingspan of a blue jay is normally distributed with a mean of 39 cm and a standard deviation of 3 cm. What is the probability that a randomly chosen blue jay has a wingspan wider than 42 cm?

13. The weight of a pink salmon is normally distributed with a mean of 1.7 kg and a standard deviation of 0.1 kg. What ratio of pink salmon is heavier than 1.9 kg?

14. Assume that the random variable $S \sim N(70, 10^2)$ describes the grades on a math test. What is the probability that a student scored more than 85 points?

15. Assume that the random variable $I \sim N(100, 15^2)$ gives the distribution of IQ test scores. What is the probability that someone's IQ is more than 120?

16. Suppose that the weight of an animal is normally distributed with a mean of 4.5 kg and a standard deviation of 2.5 kg. What is the probability that a randomly chosen animal weighs between 6 kg and 8 kg?

17. The full running speed (km/h) of a moose is normally distributed, $S \sim N(44, 5^2)$. What percent of moose can run faster than 50 km/h?

▽ 18–23 ▪ Assume that a population is normally distributed with mean μ and variance σ^2. Find the fraction of the population that falls within each interval.

18. $(\mu, \mu + 3\sigma)$

19. $(\mu - \sigma, \mu + 2\sigma)$

20. (μ, ∞)

21. $(-\infty, \mu + \sigma)$

22. $(\mu + \sigma, \infty)$

23. $(-\infty, \mu - \sigma)$

24. Assume that $X \sim N(\mu, \sigma^2)$ and let $Z = (X - \mu)/\sigma$. Using the properties of the expected value and the variance (without integrals), show that $E(Z) = 0$ and $\text{var}(Z) = 1$.

25. Assume that $X_1 \sim N(2, 12^2)$ and $X_2 \sim N(4, 6^2)$ are independent random variables. What is the distribution of $X = X_1 + X_2$? Find the mean and the variance of X.

26. Assume that $X_1 \sim N(2, 12^2)$. Under what conditions is the random variable $X = 5X_1$ normally distributed? Find the mean and the variance of X.

▽ 27–30 ▪ Suppose that $X \sim N(2, 12^2)$. Use Table 14.4 to find an x that satisfies each condition (if you cannot find an exact match, use the nearest approximation).

27. $P(X \leq x) = 0.56$

28. $P(X \leq x) = 0.95$

29. $P(X > x) = 0.2$

30. $P(X > x) = 0.3$

▽ 31–34 ▪ The grades on a math test are normally distributed with a mean grade of 72 out of 100 and a standard deviation of 8.

31. What ratio of students scored more than 90% on the test?

32. What ratio of students scored between 60 and 80 points?

33. What is the minimum score of the highest 5% of the test scores?

34. What is the maximum score of the lowest 10% of the tests scores?

35. The probability that a tree is infested by canker-rot fungus (which causes heartwood decay) is 1.4%. What is the probability that of the 200 trees selected, fewer than 25 are infested with the fungus? [Hint: Use the Central Limit Theorem.]

36. About 5% of people are carriers of a certain gene. What is the probability that in a sample of 80 people, between 5 and 10 are carriers of the gene? [Hint: Use the Central Limit Theorem.]

37. A bacterial culture contains 10,000 bacteria. Every day, each bacterium produces two offspring. The chance that they both survive is 15%, the chance that one survives is 75%, and the chance that none survive is 10%. What is the probability that the population will be larger than 10,000 tomorrow?

38. Consider the bacterial culture from Exercise 37. Find the probability that there will be between 9,000 and 12,000 bacteria the following day.

39. (Example 14.3) Use the binomial distribution $b(k, n; p)$ (with appropriate choices for k, n, and p) to find an expression for the probability that the virus will be present in between 30 and 36 months during a 10-year period. What is the major difficulty you would encounter if you tried to find the numerical value of your expression?

40. Using the substitution $u = \frac{x-\mu}{\sigma\sqrt{2}}$, show that

$$\int_{-\infty}^{\infty} \frac{1}{\sigma\sqrt{2\pi}} e^{-(x-\mu)^2/2\sigma^2}\, dx = \frac{1}{\sqrt{\pi}} \int_{-\infty}^{\infty} e^{-u^2}\, du$$

and then use (14.3) to conclude that (14.2) is true.

41. Using integration by substitution, show that

$$\int_{0}^{\infty} u e^{-u^2}\, du = \frac{1}{2}$$

Verify that the integrand is an odd function, and conclude that

$$\int_{-\infty}^{\infty} u e^{-u^2}\, du = 0$$

42. In this exercise we evaluate the integral

$$\mathrm{var}(X) = \frac{1}{\sigma\sqrt{2\pi}} \int_{-\infty}^{\infty} (x-\mu)^2 e^{-(x-\mu)^2/2\sigma^2}\, dx$$

   (a) Using integration by parts, show that

$$\int w^2 e^{-w^2}\, dw = -\frac{1}{2} w e^{-w^2} + \frac{1}{2} \int e^{-w^2}\, dw$$

   (b) Using (14.3), show that the improper integral

$$\int_{-\infty}^{\infty} w^2 e^{-w^2}\, dw$$

   is equal to $\sqrt{\pi}/2$. [You will need L'Hôpital's rule.]

   (c) Rewrite

$$\mathrm{var}(X) = \int_{-\infty}^{\infty} \left(\frac{x-\mu}{\sigma\sqrt{2}}\right)^2 \frac{\sqrt{2}\sigma}{\sqrt{\pi}} e^{-(x-\mu)^2/2\sigma^2}\, dx$$

   and substitute $w = (x-\mu)/\sigma\sqrt{2}$ to get

$$\mathrm{var}(X) = \frac{2\sigma^2}{\sqrt{\pi}} \int_{-\infty}^{\infty} w^2 e^{-w^2}\, dx$$

   (d) Use (b) to show that $\mathrm{var}(X) = \sigma^2$.

43. Prove the following for the function $g(x) = e^{-x^2}$.

   (a) $g(x)$ is an even function (i.e., symmetric with respect to the y-axis).

   (b) $g(x)$ is increasing for $x < 0$ and decreasing for $x > 0$. It has a local (also global) maximum value of 1 at $x = 0$.

   (c) The inflection points of $g(x)$ are at $x = \pm 1/\sqrt{2}$.

   (d) $\lim_{x\to-\infty} g(x) = \lim_{x\to\infty} g(x) = 0$

44. Explain how to obtain the graph of the function $f(x) = \frac{1}{\sigma\sqrt{2\pi}} e^{-(x-\mu)^2/2\sigma^2}$ from the graph of the function $g(x) = e^{-x^2}$ by shifting and scaling. Combine what you discovered with the results of Exercise 43 to prove (a)–(d) from Theorem 14. [Hint: To discover how to transform $g(x)$ to obtain $f(x)$, rewrite the exponent of e as $-\left(\frac{1}{\sigma\sqrt{2}}(x-\mu)\right)^2$.]

45. Using formula (14.10), prove the claims about the "68-95-99.7 rule" that we made in (14.9).

| 15 | The Uniform and the Exponential Distributions |

In this section we explore continuous distributions. The **uniform distribution** is the simplest of all, having a constant function for its probability density function. As the continuous analogue of the geometric distribution, we study the **exponential distribution.**

The Uniform Distribution

We have already mentioned the uniform distribution in Section 13. Now we say a bit more about it.

Definition 41   The Uniform Distribution

A continuous random variable U is said to be *distributed uniformly over the interval* $[a, b]$ in $\mathbb{R}$ if its probability density function is given by

$$f(x) = \begin{cases} \dfrac{1}{b-a} & \text{if } a \leq x \leq b \\ 0 & \text{otherwise} \end{cases}$$

Sometimes, the uniform distribution is taken to be $1/(b-a)$ on the open interval (a, b). (This makes no difference to us.)

Clearly, $f(x) \geq 0$ for all x in $\mathbb{R}$. Since

$$\int_{-\infty}^{\infty} f(x)\,dx = \int_{a}^{b} \frac{1}{b-a}\,dx = \frac{1}{b-a}(b-a) = 1$$

it follows that $f(x)$ is indeed a probability density function. The integral we calculated is the area of the shaded rectangle in Figure 15.1a.

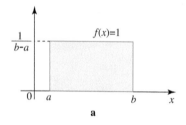
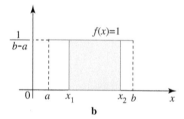

FIGURE 15.1

Uniform distribution

If $x_1, x_2 \in [a, b]$, then

$$P(x_1 \leq U \leq x_2) = \int_{x_1}^{x_2} \frac{1}{b-a}\,dx = \frac{x_2 - x_1}{b-a} \tag{15.1}$$

is the area of the rectangle over $[x_1, x_2]$ of height $1/(b-a)$; see Figure 15.1b.

The probability $P(x_1 \leq U \leq x_2)$ depends only on the difference $x_2 - x_1$, which is the length of the interval $[x_1, x_2]$. Thus, all intervals of the same length in $[a, b]$ are equally likely (hence the word "uniform").

Example 15.1   Random Numbers

The uniform distribution over the interval $[0, 1]$ is given by

$$f(x) = \begin{cases} 1 & \text{if } 0 \leq x \leq 1 \\ 0 & \text{otherwise} \end{cases}$$

Consider intervals within $[0, 1]$ whose length is 10^{-5}. According to what we just said, the intervals

$$[0, 0.00001), [0.00001, 0.00002), [0.00002, 0.00003), \ldots, [0.99999, 1) \quad (15.2)$$

are equally likely to occur.

Assume that a computer can handle numbers with up to five decimal places only (thus, for instance, it cannot distinguish between 0.345671 and 0.345674). The computer thinks that each of the intervals in (15.2) contains only one number (namely, the left endpoint).

Because the distribution is uniform, picking any of those numbers (or the corresponding intervals) is as likely as picking any other number (interval). We say that all these numbers are equally likely to occur. We know that the probability that a continuous random variable is exactly *equal* to a real number is zero. When we say that a random number generator picks any number from the interval $[0, 1)$ with equal chance, we mean it in the sense explained above. (This explains the principle; in practice, what we said needs to be modified a bit so that — as is common — a random number generator picks numbers either from a closed interval $[0, 1]$ or (more often) from an open interval $(0, 1)$.)

Suppose that a computer knows how to pick random numbers from the interval $(0, 1)$. How can you use it to generate random numbers from the interval $(3, 8)$? From (a, b)? (See Exercise 17).

Example 15.2   **Simulating Probability Using a Random Number Generator**

In Example 2.7 in Section 2 we modelled the behaviour of a molecule that, during each time interval, has a 15% chance of leaving a given region.

To simulate the probability, we used a random number generator on $[0, 1]$ and defined the simulation to run in the following way. The computer picks a random number x in $[0, 1]$. If $x \leq 0.15$, then the molecule leaves the region, and if $x > 0.15$, then the molecule stays in the region. This works because the probability of picking a number less than or equal to 0.15 is

$$P(0 \leq U \leq 0.15) = \frac{0.15 - 0}{1 - 0} = 0.15$$

(We used (15.1) with $a = 0$, $b = 1$, $x_1 = 0$, and $x_2 = 0.15$.)

Let U be a uniform distribution on $[a, b]$. Recall that its probability density function is $1/(b - a)$ if $a \leq x \leq b$ and 0 otherwise. The expected value of U is

$$E(U) = \int_{-\infty}^{\infty} x f(x) \, dx$$

$$= \int_{a}^{b} x \, \frac{1}{b - a} \, dx$$

$$= \frac{1}{b - a} \, \frac{x^2}{2} \Big|_{a}^{b}$$

$$= \frac{1}{b - a} \, \frac{b^2 - a^2}{2}$$

$$= \frac{a + b}{2}$$

since $b^2 - a^2 = (b - a)(b + a)$. Thus, the mean is (not surprisingly) the midpoint of the interval $[a, b]$.

The median of U is $(a + b)/2$ as well: the vertical line that goes through the midpoint of $[a, b]$ divides the rectangle over $[a, b]$ into two rectangles of equal area. The uniform distribution does not have a mode (or, equivalently, we can say that every number is the mode).

To calculate the variance, we start with

$$E(U^2) = \int_a^b x^2 \frac{1}{b-a}\, dx = \frac{1}{b-a} \left.\frac{x^3}{3}\right|_a^b = \frac{1}{b-a}\frac{b^3-a^3}{3} = \frac{b^2+ab+a^2}{3}$$

since $b^3 - a^3 = (b-a)(b^2+ab+a^2)$. The variance is (after simplifying fractions)

$$\mathrm{var}(U) = E(U^2) - (E(U))^2$$
$$= \frac{b^2+ab+a^2}{3} - \left(\frac{a+b}{2}\right)^2 = \frac{(b-a)^2}{12}$$

The Exponential Distribution

Recall that the geometric distribution measures the number of trials (or the time, measured in discrete intervals) until the first success. The exponential distribution is a continuous-time analogue of the geometric distribution. It measures the *exact* time when the first success occurs.

Getting the probability mass function for the geometric distribution was, more or less, straightforward. Deriving the probability density function for the exponential distribution, on the other hand, is more challenging (for completeness, we discuss it in the Appendix at the end of this section).

To start, we define a concept that is analogous to the rate of change in calculus.

Definition 42 Probabilistic Rate

If $p(\Delta t)$ is the probability of an event occurring during the time interval Δt, then the *probabilistic rate* λ is given by

$$\lambda = \lim_{\Delta t \to 0} \frac{p(\Delta t)}{\Delta t}$$

provided that the limit exists.

The units of the probabilistic rate are 1/time. For small Δt, $\lambda \approx p(\Delta t)/\Delta t$, i.e.,

$$p(\Delta t) \approx \lambda \Delta t$$

For instance, if the rate is $\lambda = 5/\mathrm{second}$, then the approximate probability of the event occurring in $\Delta t = 0.01$ seconds is $p = 5 \cdot 0.01 = 0.05$.

The time interval must be small. If, say, $\Delta t = 0.3$, then $p \approx \lambda \Delta t = 5 \cdot 0.3 = 1.5$ cannot represent a probability.

Theorem 18 The Exponential Distribution

Assume that an event occurs at a constant probabilistic rate of λ, and let T be the random variable that measures the time until the first event occurs. The probability density function of T is given by

$$f(t) = \lambda e^{-\lambda t}$$

for $t \geq 0$. The cumulative distribution function is given by

$$F(t) = 1 - e^{-\lambda t}$$

for $t \geq 0$.

We discuss the proof of this theorem in the Appendix.

Definition 43 The Exponential Distribution

The random variable T from Theorem 18 is said to be *exponentially distributed with parameter* λ.

Example 15.3   Molecular Diffusion, Continuous Case

Consider the continuous-time analogue of the molecule that diffuses out of a given region during a fixed-length time interval (see, for instance, Example 11.2 in Section 11). Suppose that the molecule leaves the region with the probabilistic rate of $\lambda = 0.2$/hour and denote by T the exact time when it leaves the region. The probability density function of T is given by

$$f(t) = 0.2\, e^{-0.2t}$$

(Figure 15.2a) and its cumulative distribution function is

$$F(t) = 1 - e^{-0.2t}$$

(Figure 15.2b).

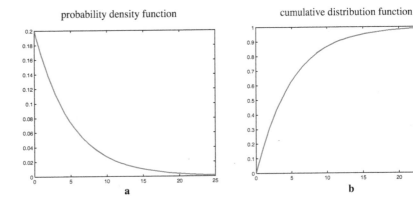

FIGURE 15.2

Exponential distribution for molecular diffusion

From

$$F(t) = P(T \le t) = 1 - e^{-0.2t}$$

we get the probability

$$P(T > t) = e^{-0.2t}$$

that the molecule is still inside the region.

The probability that a molecule leaves between 2 hours and 4 hours after the start of the experiment is

$$
\begin{aligned}
P(2 \le T \le 4) &= \int_2^4 0.2\, e^{-0.2t}\, dt \\
&= \left(-e^{-0.2t} \right)\Big|_2^4 \\
&= \left(-e^{-0.2(4)} \right) - \left(-e^{-0.2(2)} \right) \\
&= -e^{-0.8} + e^{-0.4} \approx 0.220991
\end{aligned}
$$

This probability is equal to the area of the shaded region in Figure 15.3.

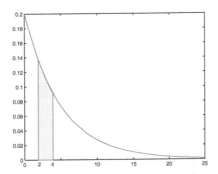

FIGURE 15.3

Probability as area for exponential distribution

Instead of integrating, we could have used the cumulative distribution function:

$$P(2 \leq T \leq 4) = F(4) - F(2)$$
$$= \left(1 - e^{-0.2(4)}\right) - \left(1 - e^{-0.2(2)}\right)$$
$$= -e^{-0.8} + e^{-0.4} \approx 0.220991$$

Assume that the lifespan of an organism is modelled by the exponential distribution with parameter λ. In that case, the cumulative distribution function is

$$F(t) = P(T \leq t) = 1 - e^{-\lambda t}$$

The *survivorship function*, i.e., the probability of being alive at time t, is given by

$$s(t) = P(T > t)$$
$$= 1 - P(T \leq t)$$
$$= 1 - F(t)$$
$$= 1 - (1 - e^{-\lambda t}) = e^{-\lambda t} \tag{15.3}$$

Example 15.4 Lifetime of a Cat

To an extent, we can model the lifetime of an animal (or a cell, or a plant) using the exponential distribution.

Assume that the lifetime of a cat is distributed exponentially with a probabilistic rate of $\lambda = 1/8$ year^{-1}.

(a) Find the probability that the cat will live longer than 15 years.

(b) Assume that the cat is 10 years old. What is the probability that it will live for at least another 15 years?

▶ Let T be the random variable that measures the lifetime of the cat. The cumulative distribution function is $F(t) = 1 - e^{-t/8}$, and $s(t) = e^{-t/8}$ is the survivorship function.

(a) The probability that the cat will live longer than 15 years is

$$P(T > 15) = s(15) = e^{-15/8} \approx 0.153355$$

So the cat has a bit over a 15% chance of living longer than 15 years.

(b) This is a question about conditional probability. We know that the cat has already lived for 10 years, and so we need to find the probability that $T > 25$ given that $T > 10$, i.e., $P(T > 25 \mid T > 10)$. Using the formula for the conditional probability (see Definition 12 in Section 4), we get

$$P(T > 25 \mid T > 10) = \frac{P((T > 25) \cap (T > 10))}{P(T > 10)}$$

The intersection of the events $T > 25$ (cat lives longer than 25 years) and $T > 10$ (cat lives longer than 10 years) is $T > 25$. Thus,

$$P(T > 25 \mid T > 10) = \frac{P(T > 25)}{P(T > 10)}$$
$$= \frac{s(25)}{s(10)}$$
$$= \frac{e^{-25/8}}{e^{-10/8}} = e^{-15/8} \approx 0.153355$$

Note that the probabilities in (a) and (b) are equal. It is certainly not realistic to expect that a cat that has lived for 10 years has the same chance of living at least 15 more years as a newborn kitten.

Thus, the exponential distribution is not appropriate as a model for all aspects of a cat's lifetime. But it is good enough as an approximation for certain aspects (such as the one we discussed in part (a)) to be still considered useful.   ◢◣

The property that the probabilities in (a) and (b) in Example 15.4 are equal is called the *non-aging property*. Interpreted for short time intervals, it means that the probability that an individual lives a bit longer is independent of its age. Or (using complementary events) the probability that an individual dies during a short interval of time is independent of its age. This kind of situation occurs with species that are lot more likely to be killed by random events than to die of old age (such as insects).

Example 15.5 **Aftershocks of an Earthquake**

Assume that the time between successive aftershocks of an earthquake is exponentially distributed with parameter $\lambda = 0.45/\text{day}$. An aftershock just occurred. What is the probability that there will be no aftershocks for more than 2 days?

▶ Denote by T the time until the next aftershock. It is given that T is exponentially distributed; its cumulative distribution function is

$$F(t) = 1 - e^{-\lambda t} = 1 - e^{-0.45t}$$

The probability that there will be no aftershocks for more than 2 days is

$$P(T > 2) = 1 - P(T \leq 2) = 1 - F(2) = e^{-0.45(2)} \approx 0.406570$$

i.e., about 41%. Note that we could have made this calculation a bit shorter by using the survivorship function $s(t) = e^{-0.45t}$:

$$P(T > 2) = s(2) = e^{-0.45(2)} \approx 0.406570$$ ◢◣

Assume that T is exponentially distributed with parameter λ. Using calculus, we can prove that

$$E(T) = \int_0^\infty t f(t)\, dt = \int_0^\infty t \lambda e^{-\lambda t}\, dt = \frac{1}{\lambda}$$

See Exercise 18(a). In Exercise 18(b) we discuss the details of the calculation

$$\text{var}(T) = \int_0^\infty \left(t - \frac{1}{\lambda}\right)^2 f(t)\, dt = \int_0^\infty \left(t - \frac{1}{\lambda}\right)^2 \lambda e^{-\lambda t}\, dt = \frac{1}{\lambda^2}$$

Thus, the mean of an exponential distribution is equal to its standard deviation.

An illustration: the average time between aftershocks in Example 15.5 is

$$E(T) = \frac{1}{0.45} \approx 2.22$$

days.

Example 15.6 **Lifetime of an Insect**

Assume that $s(t) = e^{-0.8t}$ is the survivorship function of an insect, where the time is measured in months (thus, $\lambda = 0.8/\text{month}$).

The expected value (the expected time of being alive) for the insect is $1/\lambda = 1/0.8 = 1.25$ months. We compute

$$s(1) = e^{-0.8} \approx 0.449329$$
$$s(2) = e^{-1.6} \approx 0.201897$$
$$s(3) = e^{-2.4} \approx 0.090718$$

Thus, about 45% of the insects will survive the first month, and only about 9% will survive the third month.   ◢◣

The median of the exponentially distributed random variable T is the value of t such that

$$F(t) = 0.5$$
$$1 - e^{-\lambda t} = 0.5$$
$$e^{-\lambda t} = 0.5$$
$$-\lambda t = \ln 0.5$$
$$t = -\frac{\ln 0.5}{\lambda} = \frac{\ln 2}{\lambda}$$

(since $\ln 0.5 = \ln 2^{-1} = -\ln 2$). Look familiar? See Exercise 19.

Appendix: Probability Density Function for the Exponential Distribution

We now prove Theorem 18.

In all calculations that involve the probabilistic rate we assume that Δt is small enough to guarantee that the quantities representing probabilities make sense (i.e., are smaller than or equal to 1).

At time $t = 0$ we start an experiment and measure the time until the first success. Denote by $p(t)$ the probability that, at time t, the success has not occurred yet (to make it shorter, we say that at time t the experiment is a no-success). As in calculus, we will compute

$$\frac{p(t + \Delta t) - p(t)}{\Delta t}$$

and take the limit.

The term $p(t + \Delta t)$ is the probability that the experiment is a no-success at $t + \Delta t$. It is equal to

$$p(t + \Delta t) = (\text{probability of no-success at time } t)$$
$$\cdot (\text{probability of no-success during } \Delta t)$$
$$= p(t) \cdot (\text{probability of no-success during } \Delta t) \qquad (15.4)$$

The probabilistic rate of success is λ, and thus the probability of success during the time interval Δt is $\lambda \Delta t$. Equation (15.4) implies that

$$p(t + \Delta t) = p(t)(1 - \lambda \Delta t)$$

Rearrange the terms as follows:

$$p(t + \Delta t) = p(t) - \lambda \Delta t p(t)$$
$$\frac{p(t + \Delta t) - p(t)}{\Delta t} = -\lambda p(t)$$

In the limit, as $\Delta t \to 0$, we obtain

$$p'(t) = -\lambda p(t) \qquad (15.5)$$

We recognize (15.5) as the basic exponential decay ($\lambda > 0$) differential equation. The solution is

$$p(t) = Ce^{-\lambda t}$$

where $C = p(0)$. Since the initial state of the experiment is a no-success, $p(0) = 1$ and so $C = 1$. The expression for the probability is

$$p(t) = e^{-\lambda t}$$

Let T be the exact time when the first success occurs. Then $P(T > t) = p(t)$, since success has not occurred up to and including t. It follows that the cumulative distribution function $F(t)$ is given by

$$F(t) = P(T \leq t) = 1 - P(T > t) = 1 - p(t) = 1 - e^{-\lambda t}$$

By the Fundamental Theorem of Calculus, the probability density function of T is $f(t) = F'(t) = \lambda e^{-\lambda t}$.

Summary   The probability density function of the **uniform distribution** is a constant function. Using the uniform distribution, we can explain random number generation. The exact time until the first success occurs is described with the **exponential distribution.** In certain situations, we can model the lifetime of an organism with the help of the exponential distribution.

15	Exercises

1. A uniformly distributed random variable on the interval $[0, b]$ has a variance of 12. What is its mean?

2. A uniformly distributed random variable on the interval $[a, 10]$ has a mean of 1. What is its variance?

▽ 3–8 ▪ Assume that X is an exponentially distributed random variable.

   (a) Find each probability.

   (b) Make a sketch of the probability density function and shade the region whose area is equal to the probability from (a).

3. Events occur at a constant rate of 0.2 per second. Find the probability that the first event occurs between times 2 and 6.

4. Events occur at a constant rate of 3 per second. Find the probability that the first event occurs between times 0.1 and 0.9.

5. Events occur at a constant rate of 1.5 per second. Find the probability that the first event occurs before $t = 3$.

6. Events occur at a constant rate of 1.5 per second. Find the probability that the first event occurs after $t = 4$.

7. Events occur at a constant rate of 2.4 per second. Find the probability that the first event occurs before $t = 0.3$ or after $t = 1.2$.

8. Events occur at a constant rate of 1.5 per second. Find the probability that the first event occurs before $t = 2$ or after $t = 4$.

9. Suppose that the survivorship curve for an insect is given by $s(t) = e^{-0.4t}$, where t is measured in months. What is the mean lifetime of an insect? What ratio of insects will survive 3 months?

10. Suppose that the survivorship curve for a mosquito is given by $s(t) = e^{-0.2t}$, where t is measured in months. What is the mean lifetime of a mosquito? What ratio of mosquitoes will survive 8 months?

11. Suppose that the lifetime of a radioactive atom is exponentially distributed with an expected lifespan of 4 hours. Find the probability that the atom will not decay during the first 3 hours. Find the probability that the atom will decay after 6 hours.

12. The lifetime of a radioactive atom is known to be exponentially distributed with an expected lifespan of 12 days. Find the probability that the atom will decay between 6 days and 10 days from the moment we start observing it.

13. Assume that the lifetime of a guinea pig is distributed exponentially with a probabilistic rate of $\lambda = 0.18$/year.

   (a) What is the average lifespan of a guinea pig?

(b) What is the chance that a guinea pig will live longer than 6 years?

(c) Using conditional probability, compute the chance that a 2-year-old guinea pig will live at least another 6 years. Compare with your answer to (b).

14. The lifetime of a Dalmatian is distributed exponentially with a probabilistic rate of $\lambda = 0.077/\text{year}$.

(a) What is the average lifespan of a Dalmatian?

(b) What is the chance that a Dalmatian will live longer than 12 years?

(c) Using conditional probability, compute the chance that a 5-year-old Dalmatian will live at least another 12 years. Compare with your answer to (b).

▷ 15–16 ▪ The exponential distribution implies an exponentially decreasing survivorship curve, which means that an organism is more likely to die when young. Explain what kind of survivorship is implied by each (non-exponential) graph.

15.

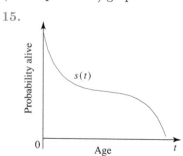

16.

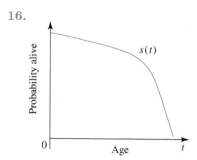

17. Find a formula for a linear function that maps the interval $(0, 1)$ to the interval $(3, 8)$. Explain how you can use this function to generate random numbers in $(3, 8)$, assuming that you can generate random numbers in $(0, 1)$. Adjust this construction so that you are able to generate random numbers from any interval (a, b).

18. We calculate the integrals that we need for the mean and the variance of the exponential distribution.

(a) Using integration by parts with $u = t$ and $v' = e^{-\lambda t}$, show that
$$\int_0^\infty t \lambda e^{-\lambda t}\, dt = \frac{1}{\lambda}$$

(b) Use integration by parts twice (start with $u = (t - 1/\lambda)^2$ and $v' = e^{-\lambda t}$) to verify that
$$\int_0^\infty \left(t - \frac{1}{\lambda}\right)^2 \lambda e^{-\lambda t}\, dt = \frac{1}{\lambda^2}$$

19. Assume that $s(t) = e^{-\lambda t}$ describes the decay of a radioactive substance. Find its half-life. Also, find the median of the exponentially distributed random variable whose cumulative distribution function is $F(t) = 1 - e^{-\lambda t}$. Explain why the half-life and the median are equal.

20. Assume that the random variable T describing the lifetime of a radioactive atom is distributed exponentially with parameter $\lambda = 30/\text{year}$. Find the time t_0 such that $P(T > t_0) = 1/2$. What is the meaning of t_0? How is it related to the median of T?